Problems and Solutions
for Strachan & Read's
Human Molecular Genetics 2

Problems and Solutions for Strachan & Read's Human Molecular Genetics 2

David J. Matthes

Department of Biological Sciences
San Jose State University
San Jose, California

WILEY-LISS

A JOHN WILEY & SONS, INC., PUBLICATION

New York • Chichester • Weinheim • Brisbane • Singapore • Toronto

Published in the United States of America, its dependent territories and Canada by John Wiley & Sons, Inc., and in Asia and its dependent territories, Australia and India by John Wiley & Sons (Asia) Pte Ltd by arrangement with BIOS Scientific Publishers Limited, 9 Newtec Place, Magdalen Road, Oxford OX4 1RE, UK.

A CIP catalogue record for this book is available from the British Library.

ISBN 0-471-38414-3

Library of Congress Cataloging in Publication Data
Strachan, T.
Human molecular genetics / Tom Strachan and Andrew P. read. — 2nd ed.
p. cm.
Includes bibliographical references and index.
ISBN 0–471–33061–2 (pbk.)
1. Human molecular genetics. I. Read, A. P. (Andrew P.)
II. Title.
QH442.S775 99–38373
6119.01816—dc21 CIP

USA
John Wiley & Sons, Inc.,
605 Third Avenue, New York,
NY 10158-0012, USA

Canada
John Wiley & Sons (Canada) Ltd,
22 Worcester Road, Rexdale,
Ontario M9W 1L1, Canada

Asia
John Wiley & Sons (Asia) Pte Ltd
2 Clementi Loop, #02-01
Singapore 129809

Australia
John Wiley & Sons Australia Ltd
33 Park Road, PO Box 1226
Milton, Queensland 4046, Australia

Production Editor: Fran Kingston
Typeset by Phoenix Photosetting, Chatham, Kent, UK
Printed by The Cromwell Press, Trowbridge, Wiltshire, UK

Contents

Abbreviations

7mG	7-methylguanosine
ARMS	amplification refractory mutation system
ASO	allele specific oligonucleotides
CCM	chemical cleavage of mismatches
CXIF	cytokine X-inducible factor
dATP	deoxyadenosine 5′-triphosphate
DDGE	denaturing gradient gel electrophoresis
DMD	Duchenne muscular dystrophy
DNA	deoxyribonucleic acid
DNase I	deoxyribonuclease I
dNTP	deoxynucleotide triphosphate
DR4	dopamine receptor 4
dsDNA	double-stranded DNA
ES	embryonic stem
EST	expressed sequence tag
FISH	fluorescent *in situ* hybridization
GTP	guanosine 5′-triphosphate
HPLC	high performance liquid chromatography
HPRT	hypoxanthine phosphoribosyltransferase
IQ	intelligence quota
LINE	long interspersed nuclear element
LOD	logarithm of the odds ratio
LTR	long terminal repeat
MAOA	monoamine oxidase A
mRNA	messenger RNA
OLA	oligonucleotide ligation assay
OMIM	Online Mendelian Inheritance in Man
ORF	open reading frame
pfu	plaque forming units
PCR	polymerase chain reaction
PGFE	pulsed field gel electrophoresis
PolyA+RNA	polyadenylated RNA
PTT	protein truncation test
RACE	rapid amplification of cDNA ends
RFLP	restriction fragment length polymorphism
RNA	ribonucleic acid
RNAse	ribonucelase
RT	reverse transcriptase
SAGE	serial anaysis of gene expression
SLE	systemic lupus erythematosus
SSCP	single-strand conformation polymorphism

STS	sequence tagged site
SVAS	supravascular aortic stenosis
TFIIA	transcription factor IIA
TFIID	transcription factor IID
TK	thymidine kinase
TMA	trimethylethylamine
TNF	tumor necrosis factor
tRNA	transfer RNA
TS	tumor suppressor
UTR	untranslated region
UV	ultraviolet
VNTR	variable number of tandem repeats
YAC	yeast artificial chromosome

Preface

This book of multiple choice questions, review questions and open-ended problems is designed to be a companion volume to *Human Molecular Genetics* 2nd edition, by Tom Strachan and Andrew Read. My objective is to provide a supplement to the textbook of students to use as part of a university course or on their own. These solved problems provide the reader with the means to test their comprehension of central concepts addressed in each chapter of the textbook. I believe that by applying what one has read to challenging problems, one acquires a deeper understanding of the material and a greater appreciation of its usefulness.

Instructors using *Human Molecular Genetics,* 2nd edition, in their courses will appreciate that the problems are not, for the most part, beyond the reach of an average molecular genetics student. The multiple choice questions are challenging and require students to possess a thorough understanding of the material before they can be answered correctly. Most questions require students to think carefully to avoid easy misconceptions. With these problems, students can quickly check their comprehension of a chapter they have read using the multiple choice questions and review questions, or they can pursue selected topics to a greater depth by focusing on the open-ended problems.

Writing problems and solutions is a time consuming endeavor. I hope that genetics instructors and teaching assistants who use this book with their students will be saved some of the time they would otherwise have spent coming up with problems of their own.

Finally, any corrections, suggestions for improvement, or suggestions of topics in Human Molecular Genetics that should have questions or problems written for them are very welcome. You can fax them to me at (408) 924 4840 or send them to BIOS Scientific Publishers who will pass them on to me.

David J. Matthes

Acknowledgements

I would like to thank Tom Strachan and Andrew Read for writing *Human Molecular Genetics*, a book as perfectly suited for advanced biology students as for professional human molecular geneticists. I also want to thank the past and present students in my molecular genetics course for their helpful comments on early versions of these problems, and for confirming for me the importance of problem-solving for the student of genetics. Thanks also to Jonathan Ray, Managing Director at BIOS, who encouraged this project from the beginning, and to Fran Kingston for her careful attention to the details of the book's production. Finally, my deepest appreciation to my wife Devavani, whose support and encouragement have made the steepest mountains seem merely hills.

1 DNA structure and gene expression

Multiple choice questions

1. The base shown at right is

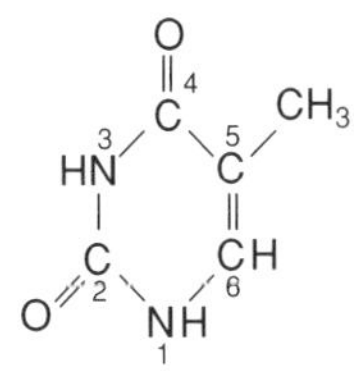

A 5-methyl cytosine
B cytosine
C thymine
D adenine
E none of the above

2. The base labeled 'hydrogen bond partner' usually pairs with which of the following when it is incorporated in dsDNA?

A

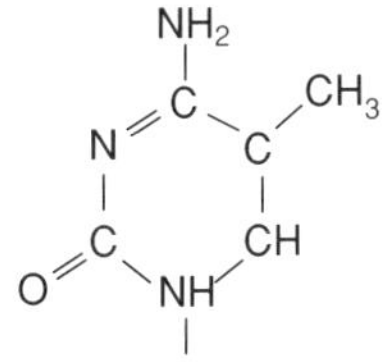

B

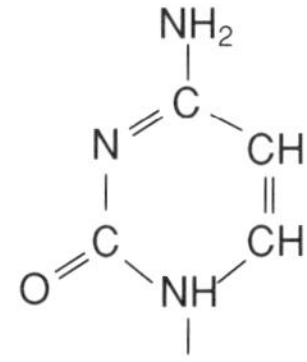

C

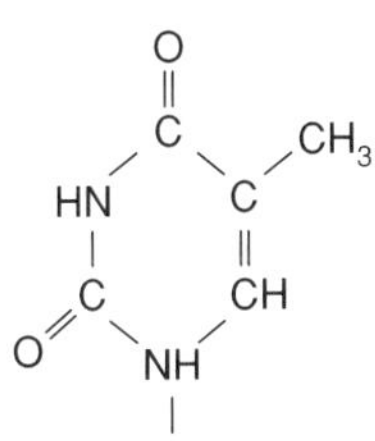

D

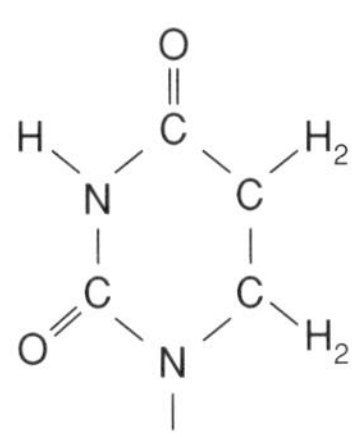

E

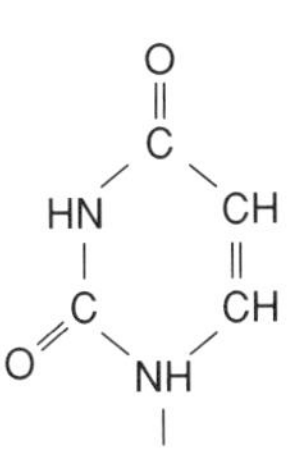

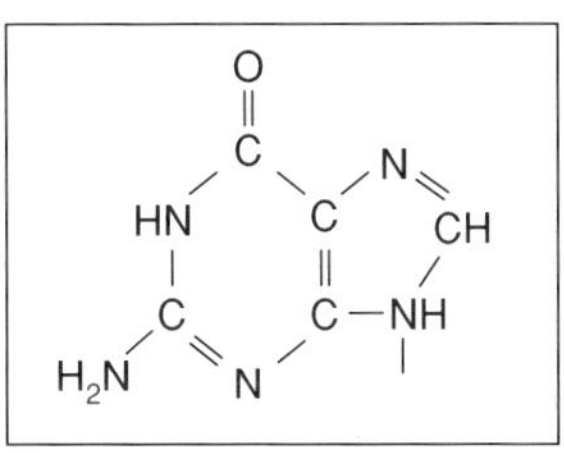

Hydrogen bond partner

3. Which of the following is a feature of DNA but not RNA?

- **A** Newly synthesized strand is anti-parallel to the template
- **B** Contains a 2′ hydroxyl group on the sugar moiety of the constituent nucleotides
- **C** New strands are synthesized in the 5′ to 3′ direction
- **D** Assumes complex secondary structures
- **E** Synthesis is semi-conservative

4. Eukaryotic mRNAs are generally all of the following EXCEPT

- **A** transcribed by RNA Polymerase II
- **B** 7mG capped, spliced to remove introns, and tailed with poly(A)
- **C** the most abundant type of RNA in a cell
- **D** lacking promoter and enhancer sequence but containing 5′ and 3′ UTRs
- **E** colinear with the proteins they encode

5. The capping of mRNA at the 5′ end with 7-methylguanosine is thought to

- **A** protect the transcript from exonuclease attack
- **B** facilitate transport from the nucleus to the cytoplasm
- **C** facilitate RNA splicing
- **D** all of the above
- **E** none of the above

6. Which of the following best represents the intronic splice product as it would be immediately after splicing of the fragment of mRNA shown below? Note: the line indicates which bases are covalently bound to form a lariat structure.

5′ UACCUGAAGGUAAGUGUUCCUAGUCACAUCAACGGGCUUCCUCCUCCUGCAGGCUG 3′

- **A** 5′ GUAAGUGUUCCUAGUCACAUCAACGGGCUUCCUCCUCCUGCAG 3′
- **B** 5′ UACCUGAAGGCUG 3′
- **C** 5′ GUAAGUGUUCCUAG 3′
- **D** 5′ CACAUCAACGGGCUUCCUCCUCCUGCAG 3′
- **E** 5′ UACCUGAAGGUAAGUGUUCCUACUCCUCCUGCAGGCUG 3′

7. How would the following mRNA be translated by eukaryotic ribosomes? You may assume the transcript has already undergone post-transcriptional processing.

5′ 7mG A A G C C C A U G A C C A U G G A A U U C U G A U G U A A . . . (poly A) 3′

Optimal site of initiation of translation: 5′ . . . G C C <u>**A**</u> C C <u>A U G</u> **G** . . . 3′ (**A** or **G**)

- **A** $^{+}NH_3$ E A H D H G I L M . COO^-
- **B** $^{+}NH_3$ M A T M E F. COO^-
- **C** $^{+}NH_3$ K P M T M E F. COO^-
- **D** $^{+}NH_3$ M E F. COO^-
- **E** $^{+}NH_3$ M . COO^-

One-letter codes for amino acids

A alanine	**C** cysteine	**D** aspartic acid	**E** glutamic acid	**F** phenylalanine
G glycine	**H** histidine	**I** isoleucine	**K** lysine	**L** leucine
M methionine	**N** asparagine	**P** proline	**Q** glutamine	**R** arginine
S serine	**T** threonine	**V** valine	**W** tryptophan	**Y** tyrosine

8. Which of the following is an uncharged polar amino acid that in glycoproteins can be the site of carbohydrate attachment in the context 'N X S/T'?

A Aspartate
B Asparagine
C Alanine
D Cysteine
E None of the above

9. Which of the following is NOT a major type of post-translational modification?

A Methylation of lysine
B Hydroxylation of proline
C Carboxylation of glutamate
D Methylation of cysteine
E Glycosylation of serine

10. The secondary structure of a protein is considered to be

A the overall three-dimensional structure of a polypeptide
B the linear sequence of amino acids in a polypeptide
C the path that the polypeptide backbone follows in space
D the folded arrangement of subdomains of the mature protein
E the structure that results from proteolytic cleavage of protein precursors

Review question

1. (a) Draw the two base pairs most commonly found in DNA double helices. Show all chemical groups and the main covalent and hydrogen bonds.
   (b) Number the carbons and nitrogens of the bases appropriately.
   (c) Label the site of methylation of guanine when GTP is used to cap mRNAs, the site of methylation in 5-methylcytosine, and the site of deamination of cytosine (and 5-methylcytosine).

2. Most naturally occurring proteins are composed of 20 different kinds of amino acids.

   (a) List which of the standard amino acids fall into each of the following chemical groups: acidic, basic, uncharged polar and nonpolar neutral.
   (b) Which two amino acids have the largest side groups?
   (c) Which one is most directly involved in the formation of disulfide bridges?
   (d) Which one is encoded by the sequences of bases: 5′ ATG 3′?

3. Distinguish between:

   (a) TATA box and GC box.
   (b) Promoter and enhancer.
   (c) Antiparallel and complementary (with respect to nucleic acid strands).
   (d) Splice junction and branch site.

Open-ended problems

Data

Splice donor, splice branch, and splice acceptor sites:

```
5′ C A G | G T A A G T ... T N C T A A C...  (C)    N C A G | G  3′
   A           G           C   T C G         (T)11    T       A
```

1. A genomic DNA sequence (sense strand) and protein sequence (conceptually translated from the genomic DNA in all three frames) is shown below. The boxed amino acids in the sequence correspond closely to amino acids in a protein encoded by a related organism, shown below the first sequence.

```
                                          27                                          54
GAG GAT GAT TGT CAG AAT TAC ATT AGA ATA CTG TAC TCT TCA GAA CCG GGG AAA
E   D   D   C   Q   N   Y   I   R   I   L   Y   S   S   E   P   G   K
 R   M   I   V   R   I   T   L   E   Y   C   T   L   Q   N   R   G   N
  G   .   L   S   E   L   H   .   N   T   V   L   F   R   T   G   E   I

                                          81                                         108
TTA GTT ATT TGC GGG ACC AAT TCG TAC AAA CCC CTC TGT CGG ACG TAC GCA TTT
L   V   I   C   G   T   N   S   Y   K   P   L   C   R   T   Y   A   F
 .   L   F   A   G   P   I   R   T   N   P   S   V   G   R   T   H   L
  S   Y   L   R   D   Q   F   V   Q   T   P   L   S   D   V   R   I   .

                                         135                                         162
AAG GTA GGT CCA GTC GGT TTG ACG CAA TCG TCA TTT ACA CTC CAC ATA ATT CTA
K   V   G   P   V   G   L   T   Q   S   S   F   T   L   H   I   I   L
 R   .   V   Q   S   V   .   R   N   R   H   L   H   S   T   .   F   .
  G   R   S   S   R   F   D   A   I   V   I   Y   T   P   H   N   S   R

                                         189                                         216
GGA GGG AAA GTA CCT GGT TGA GAA AGA AGT AGA AGG GAT AGG CTT GTG TCC ATA
G   G   K   V   P   G   .   E   R   S   R   R   D   R   L   V   S   I
 E   G   K   Y   L   V   E   K   E   V   E   G   I   G   L   C   P   Y
  R   E   S   T   W   L   R   K   K   .   K   G   .   A   C   V   H   T

                                         243
CAA TCC GGA ACA CAA CAG CAC ATC TGT CTC CTA CAA TGG CCA A
Q   S   G   T   Q   Q   H   I   C   L   L   Q   W   P
 N   P   E   H   N   S   T   S   V   S   Y   N   G   Q
  I   R   N   T   T   A   H   L   S   P   T   M   A
```

The amino acid sequence of the same part of the homologous protein from a closely related species:

EDDCQNYIRVLAKIDDDRVLICGTNAYKPLCRHYAL
KDGDYVVEKEYEGRGLCPFDPDHNSTAIYSEGQ

Hint: the consensus sequences you need to know are in Human Molecular Genetics, *2nd edn.*

(a) Identify the splice donor and acceptor sequences. Underline the bases that match the consensus sequence. Put a dot over the bases at the splice sites you have chosen that don't match the consensus.
(b) Identify possible branch sites. Identify the one(s) that best match the branch site consensus sequence. Underline the bases that match the branch site consensus sequence. Put a dot over the bases that do not match the branch site consensus.
(c) Connect by a line both pairs of bases that are most likely to be joined together during lariat formation and intron removal.
(d) Box the amino acids that are conserved between the two homologous sequences. Underline the amino acids that are chemically very different. Indicate the amino acid that is most likely to be glycosylated post-translationally.

2. The double-stranded DNA sequence shown below is from the gene encoding the SP1 transcription factor. Most of the promoter sequence (the upstream sequence is numbered with negative numbers) and part of the first exon (transcription begins at +1) are shown.

```
      -321       -311       -301       -291       -281       -271
GGGGGCGGGC GGGAGGCGGC GGCGGCGGCG GCCTAGGTCC CGCCCGGGGC GGAGGGAAGG
CCCCCGCCCG CCCTCCGCCG CCGCCGCCGC CGGATCCAGG GCGGGCCCCG CCTCCCTTCC

      -261       -251       -241       -231       -221       -211
GAGGGAGACG GGGCAGTGGC GGGGCCTCCG AGGAGGAGGG GGATGGGCCG CCCGCCCCGG
CTCCCTCTGC CCCGTCACCG CCCCGGAGGC TCCTCCTCCC CCTACCCGGC GGGCGGGGCC
      -201       -191       -181       -171       -161       -151
GGGAGGGGGC AGCGTGGCCT CGCCCGCCCC CTGCCCGCCC CGGCCACGGG GGACGGGCCT
CCCTCCCCCG TCGCACCGCA GCCGTCACCG CCCCGGAGGC TCCTCCTCCC CCTGCCCGGA

      -141       -131       -121       -111       -101        -91
TACCCCCCAC TACTCGGCCG CCCGCCTGAG GCTCCTCCCG CCGGGGGCTG GAGCCGCGGG
ATGGGGGGTG ATGAGCCGGC GGGCGGACTC CGAGGAGGGC GGCCCCCGAC CTCGGCGCCC

       -81        -71        -61        -51        -41        -31
GGCGGCCCGA GCAGCGAAGG CCCCGCCCGG GCCAACCGCC TGCCTGGTCC GCCCTCTGGT
CCGCCGGGCT CGTCGCTTCC GGGGCGGGCC CGGTTGGCGG ACGGACCAGG CGGGAGACCA

       -21        -11         -1         10         20         30
CGCCGCCTGC TCCAAGGCCC TCCTCCCCCC ACTTTCCGTA GATTTCCCTT CCCCCCCGCC
GCGGCGGACG AGGTTCCGGG AGGAGGGGGG TGAAAGGCAT CTAAAGGGAA GGGGGGGCGG

        40         50         60         70         80         90
CCCCACCGTC CCGCCCTTTC CACGCCCCTC ACCCCGAAAC CGCCCTTTCC CTCAGGGCCC
GGGGTGGCAG GGCGGGAAAG GTGCGGGGAG TGGGGCTTTG GCGGGAAAGG GAGTCCCGGG
```

(a) Write out the first 10 bases of the primary RNA transcript of the SP1 gene.
(b) How would this 5′ end be different in the mature messenger RNA?
(c) What evidence is there that the SP1 transcription factor might, in part, regu-

late its own transcription? (Hint: the sequences marked with a dark black line indicate regions of the gene that are 'protected', or covered by proteins, when nuclear proteins are added to this DNA.)

(d) Does this gene use a TATA box in its promoter? How does your answer reconcile with the fact that SP1 is a transcription factor that regulates constitutively expressed genes?

2 Chromosomes in cells

Multiple choice questions

1. During the eukaryotic cell cycle the genomic DNA content, measured in units of C, for a given cell

A stays constant from the end of G_2 to the beginning of G_1
B doubles during mitosis
C stays constant in interphase
D halves upon cytokinesis
E doubles when ploidy doubles, halves when ploidy halves

2. X chromosomes are all of the following EXCEPT

A inactivated by lyonization in females with Turner's syndrome (45,X)
B inactivated at random in cells early in female embryogenesis
C inactivated in females, and preserved that way in the descendants of a cell once the initial inactivation has occurred
D able to pair with a Y chromosome at the major pseudoautosomal region
E obliged to cross-over with the Y chromosome during male meiosis

3. Packaging of DNA into chromosomes requires multiple hierarchies of DNA folding, mediated by proteins at many steps. Which pairing of level of compaction with structural protein involved in generating the structure is NOT correct?

A 30 nm chromatin fibers require histone 1
B Nucleosomes require histones 2A, 2B, 3 and 4
C Chromatin loops require scaffold proteins
D Double helices require helicase
E None of the above

4. All of the following have been proposed to be functions of telomeres EXCEPT

A to ensure complete replication of the extreme ends of chromosome termini
B to allow the segregation of eukaryotic chromosomes during mitosis
C to maintain the structural integrity of a chromosome
D to help establish the three-dimensional architecture of the nucleus
E to promote chromosome pairing

5. What nucleotides would be added by telomerase (with an internal RNA template 5′ CAACCCCAA . . . 3′) to the chromosome end below in the first cycle of DNA synthesis before enzyme translocation?

5′ . . . TTGGGGTTGGGGTTGGGGTTG 3′
3′ . . . AACCCC 5′

A 5′ CAACCCCAACCCCAA 3′
B 5′ CAACCCCAA 3′
C 5′ TTGGGGTTG 3′
D 5′ GGGTTG 3′
E 5′ CAACCC 3′

6. During gametogenesis male and female germlines mature on different developmental timelines. Which of the following is true about the development of human germlines?

A At puberty, hormones induce the completion of meiosis I in all primary oocytes which subsequently complete meiosis II in monthly intervals
B The transition from 2n to n occurs between oogonium and primary oocyte
C Secondary oocytes complete meiosis II after fertilization
D Secondary spermatocytes complete meiosis I in the fetus and complete meiosis II in monthly intervals beginning at puberty and continuing until death
E Meiotic arrest in oocytes accounts for the reduced rate of chromosomal nondisjunction in women over 30 years old

7. The model of chromosome banding proposed by Saito and Laemmli (1994) provides an explanation for the basis of the R and Q loops of metaphase chromosomes. Which of the following descriptions apply to the region of metaphase chromosome indicated by the arrow?

A AT-rich SARs are found more densely here
B This region has contains tighter DNA loops
C Giemsa stain has a particularly high affinity
D Poly(A)+ mRNA hybridizes here preferentially
E Would be stained preferentially in C-banding procedures

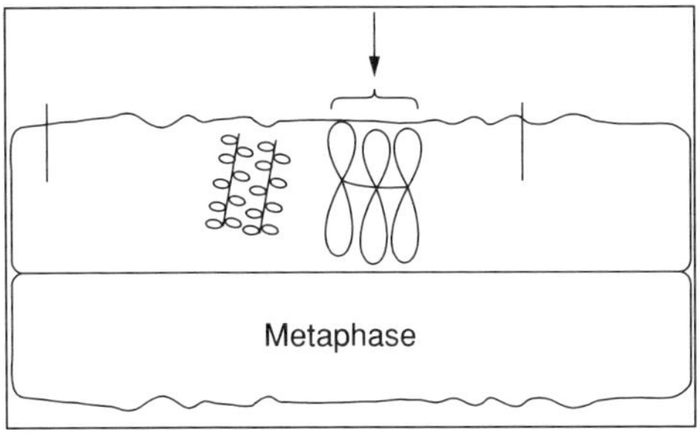

8. Which of the following chromosomes best represents the defective chromosome found in a person with the karyotype: 46, XX, inv(17) (q12q23)?

#17

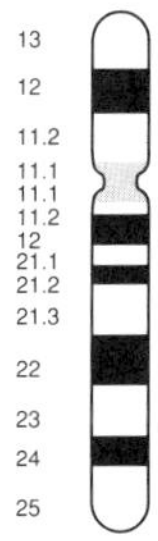

A

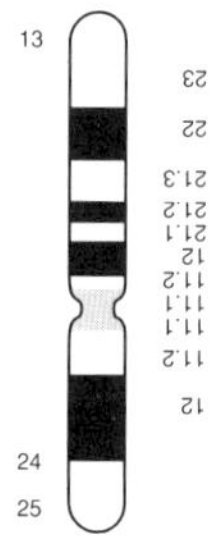

B

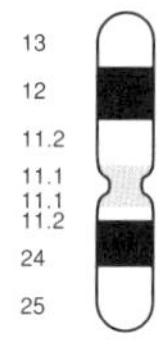

C

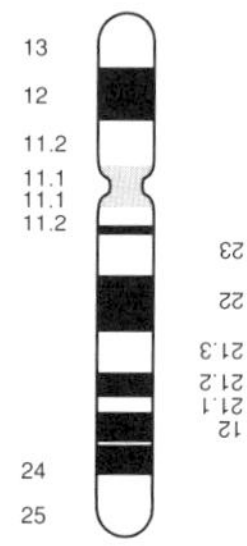

D

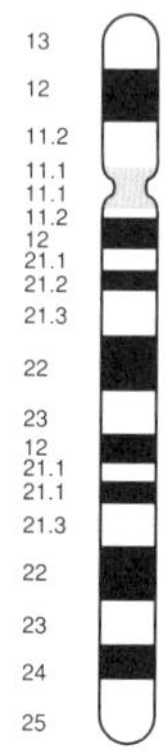

E None of these

9. Aneuploidy can result from

A prophase lag
B base substitution mutation
C frame shift mutation
D nondisjunction
E none of the above

10. Which of the following pairs of chromosomal abnormality and outcome of that abnormality is NOT correct?

A Ring chromosome : chromosome loss if ring does not include centromere
B Inversion : reduced fertility proportionate to the size of the inversion
C Balanced translocation : semi sterility
D Centric fusion : contiguous gene syndrome
E Duplication : genetic disease or lethality due to gene dosage problems

Review questions

1. Which would you expect to have more methylated DNA, regenerating hepatocytes or platelets? Why?

2. What are n and C for mouse? For onion?

3. What roles have been proposed for nucleosomes?

4. What are the three essential elements of chromosomes? Briefly describe each on a molecular level.

5. Define mixoploidy and two ways in which it can arise.

6. Explain the difference between G banding and R banding in terms of the nature of the genetic material stained in each case.

7. What is the function of each of the following procedures during the preparation of chromosomes for cytological analysis?

   (a) Phytohemagglutinin treatment
   (b) Thymidine starvation
   (c) Colcemid treatment
   (d) Hypotonic saline treatment
   (e) Trypsinization
   (f) Heat denaturation in saline
   (g) Giemsa staining
   (h) Quinacrine, DAPI, or Hoechst 33258 treatment
   (i) Chromomycin A3, olivomycin, or mitramycin treatment
   (j) Barium hydroxide treatment

Open-ended problems

1. Give three or more different ways to identify a centromere in a mammalian chromosome?

2. Sketch the relevant chromosomes for the following karyotype: 46, XX t(2;6) (q35;p21.3). You may want to shade chromosome 2-derived material for clarity.

3. Assuming all of the chromosomes that are not shown are normal, what is the karyotype of the patient possessing the chromosomes shown below? What is the abnormality?

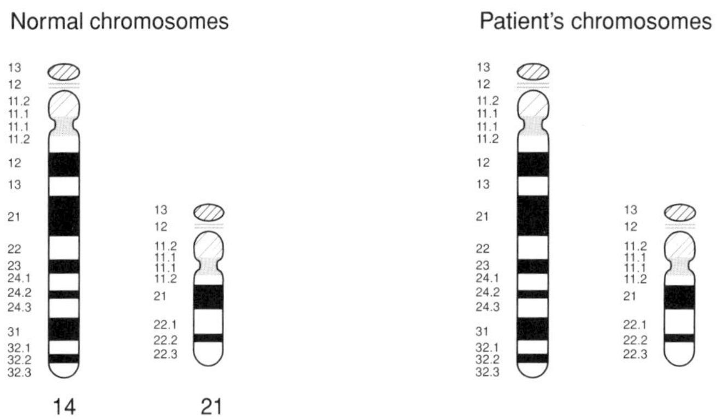

4. Many molecular biological analyses of human genomic DNA involve the isolation of human genomic DNA. As blood is a convenient source of human cells, it is often the source material for human genomic DNA. How many milliliters of blood would be required to isolate 100 micrograms of genomic DNA from lymphocytes purified from human blood? Assume that you can isolate 5000 lymphocytes from each microliter of whole blood.

5. If humans have 100 000 genes, approximately how many genes are there in a single band (on average) in a high-resolution G banding chromosome spread? How many base pairs in a single band, on average?

3 Genes in pedigrees

Multiple choice questions

1. How would one use pedigree notation to indicate the dead twins of unknown sex of a man and his cousin?

A B C D E

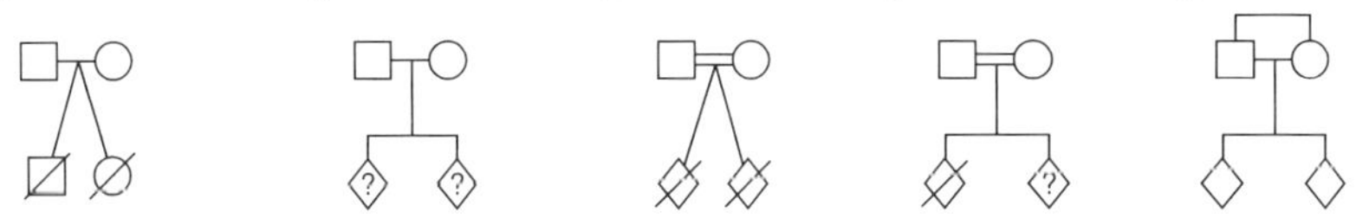

2. What is the most likely genotype of individual III_1 in the pedigree below?

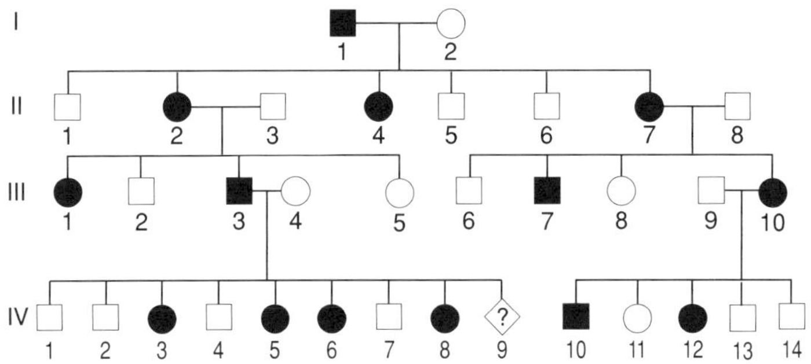

A Dd
B dd
C X^DX^d
D X^dX^d
E Mitochondrial 'D'

3. Two people each homozygous for autosomal recessive deafness-producing alleles have four children. Three are deaf. One can hear. Possible explanations for the appearance of the hearing child include all of the following EXCEPT

A complementation of two different loci (AA bb × aa BB → Aa Bb [hearing])
B back mutation (dd → Dd)
C heterozygosity at loci for which spouse is homozygous (Aa bb × aa Bb → Aa Bb)
D infidelity / mistaken parentage
E late or variable age of onset (hearing child will lose hearing later in childhood)

4. The pedigree below is consistent with all of the following modes of inheritance EXCEPT

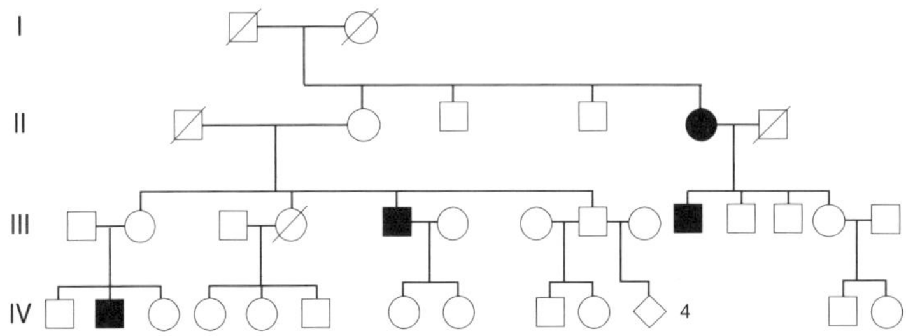

A autosomal dominant (incompletely penetrant)
B autosomal recessive (common allele)
C sex-linked dominant (incompletely penetrant)
D sex-linked recessive (common allele)
E mitochondrial (incompletely penetrant)

5. The pedigree below reveals that individual III_9

A has a new autosomal dominant mutation
B has a sex-linked recessive condition
C has an autosomal recessive condition
D exhibits penetration of an incompletely penetrant dominant allele
E could have any of the above

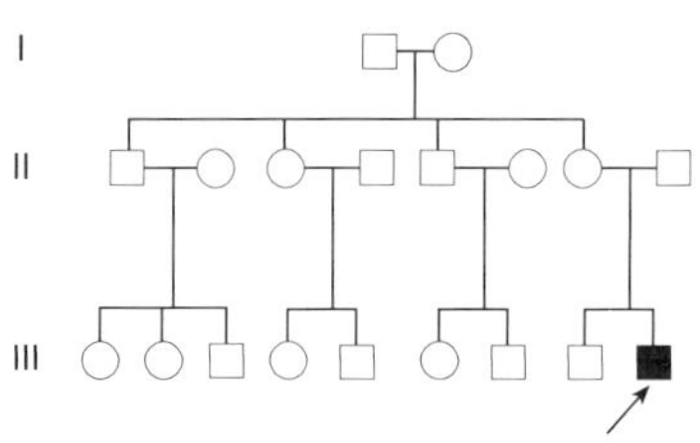

6. The maternal and paternal genomes in an individual do not always function interchangeably. Which of the following observations does NOT support that idea?

A mouse embryos with two maternal genomes fail to develop
B human triploid abortuses are phenotypically different depending on whether the extra chromosome is maternal or paternal
C allele loss in many cancers preferentially involves the paternal allele
D some human diseases are autosomal dominant but manifest only when inherited from one parent
E deletion of certain chromosomal regions produces the same phenotype whether on the paternal or maternal chromosome

7. If one were investigating the hypothesis that a gene that is perhaps be Y-linked (such as the one for the human porcupine trait), it might be informative to interview people in a community where the trait was common. Which of the following would be the least likely to yield an answer *opposed to* the Y-linkage hypothesis, if asked of an affected man?

A Do your father's brothers have the condition?
B Do all of your brothers have the condition?
C Do all your sons have the condition?
D Do your mother's brothers have the condition?
E None of the above would provide any useful data

8. Mosaicism differs from chimerism in which of the following ways?

A Mosaics are derived from one zygote while chimeras arise from the fusion of two (or more) different zygotes or embryonic stem cells
B Chimeras are genetically identical from cell to cell, except for one locus, while mosaics have multiple alleles at many loci
C All females (with a randomly inactivated X chromosome in every nucleated cell) are mosaic while all males (with an X chromosome from their mother and a Y chromosome from their father) are chimeras
D Mosaics have heterogeneous germlines while chimeras have heterogeneous somatic cells
E While one can be mosaic to varying degrees (e.g. 1/10 Aa and 9/10 AA) one cannot be chimeric to varying degrees (e.g. 1/10 A1 A2, 9/10 A3 A4)

9. If cystic fibrosis, an autosomal recessive condition, occurs in 1 in 2000 Caucasian births, what is the approximate frequency of heterozygotes in the Caucasian population?

A 1 – square root of 2000
B 999/2000
C Square root of 2000 × (1 – square root of 2000) × 2
D 1/2000 × 1999/2000 × 2
E None of the above

10. If cystic fibrosis is an autosomal recessive condition that until recently prevented the affected individual from surviving to reproductive age, what is the mutation rate, mu (in units of mutations per *CFTR* locus per generation) that would be necessary to maintain the 'c' allele frequency approximately unchanged over the preceding generations *assuming no selection pressure on heterozygotes*? 1 in 2000 Caucasians are born with cystic fibrosis.

A The square root of 1/2000
B 1/2000
C 1/4 000 000
D 0
E None of the above

Open-ended problems

1. A tremendous amount of information about human genetics is available online at the Online Mendelian Inheritance in Man (OMIM) website.

   (a) Obtain a complete listing for any Mendelian character in humans using OMIM online.
   (b) Summarize the mutant or variant phenotype(s) succinctly.
   (c) Identify the dominance/recessiveness of the character.
   (d) Mention any complexities in the inheritance pattern.
   (e) Include the first three pages of the online listing with your response.

2. Two people with autosomal recessive profound deafness marry each other and have children, all but one of whom are also deaf. Propose a reasonable hypothesis to explain these progeny.

3. Careful molecular analysis of mitochondrial DNA from a male patient reveals two different restriction fragment lengths identified with a particular probe. Only one restriction fragment length is found in the man's mother. How would you distinguish between a very rare instance of paternal mitochondrial transmission and polymorphism due to new mutation?

4 Cell-based cloning

Multiple choice questions

1. The four essential steps of cell-based DNA cloning include all of the following EXCEPT

A formation of recombinant DNA
B transformation
C primer design and target extension
D amplification to produce numerous clones
E isolation of recombinant clones

2. *Xma*III cuts at 5′ C^GGCCG 3′. *Xma*III is an example of

A a type I restriction enzyme that leaves a 5′ overhang
B a type I restriction enzyme that leaves a 3′ overhang
C a type II restriction enzyme that leaves a 5′ overhang
D a type II restriction enzyme that leaves a 3′ overhang
E an isoschizomer of *Eco*RI

3. One common way of preventing or minimizing vector-cyclization is

A vector dephosphorylation
B cutting of vector with a single restriction endonuclease
C formation of insert-insert concatamers
D ligation reaction in large volume
E insert dephosphorylation

4. Which restriction map of a 1200 bp PCR fragment best fits the data given below?

Enzyme used	Length of fragments (bp)
*Sma*I	1000, 200
*Hind*III	500, 400, 300
*Eco*RI	750, 450
*Sma*I, *Hind*III	500, 300, 200
*Hind*III, *Eco*RI	450, 400, 300, 50
*Sma*I, *Eco*RI	750, 250, 200

A 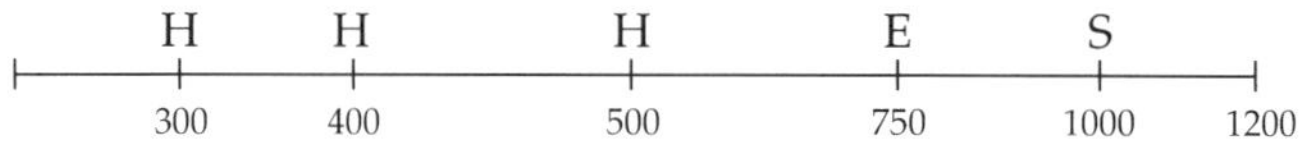

B 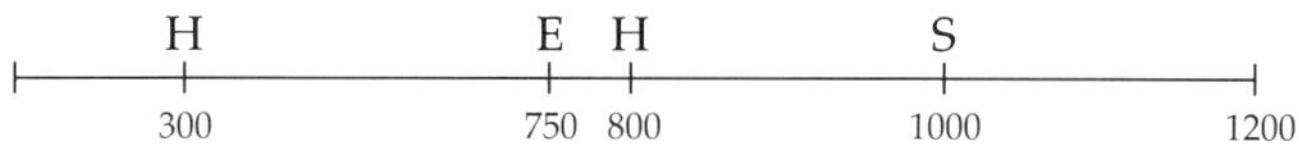

C 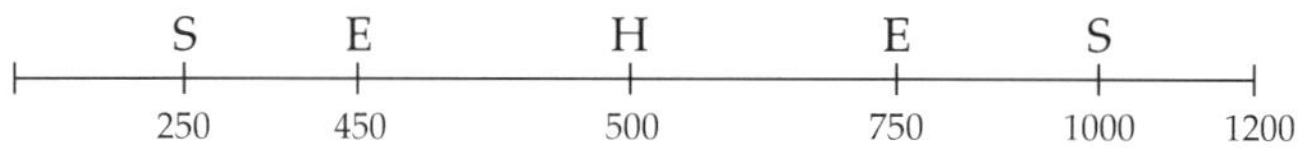

D 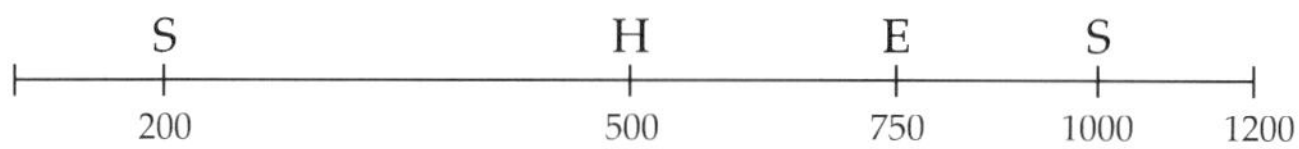

E None of the above

5. All of the following are essential components of a plasmid vector EXCEPT

A a selectable marker
B insertionally inactivatable marker
C origin of replication
D unique restriction site
E circularity

6. For making a genomic library with inserts between 30 and 40 kb in length, which of the following cloning vectors would be optimal?

A Bacteriophage P1
B YAC
C Plasmid
D Bacteriophage lambda
E Cosmid

7. Expression cloning in *E. coli* requires all of the following EXCEPT

A an antibody to a protein epitope
B cDNA inserted behind promoters recognized by *E. coli* RNA polymerase
C transfer of colonies or plaques to a stable membrane
D screening a genomic library with a probe to a gene's regulatory sequence.
E an antibody that will recognize filter-bound epitopes

8. **Which of the following vector systems uses *cre* recombinase?**

A Cosmids
B YACs
C P1s
D Plasmids
E None of the above

9. **The diagram below illustrates which of the following?**

A Genomic cloning
B Expression cloning
C Antibody production
D Affinity chromatography
E None of the above

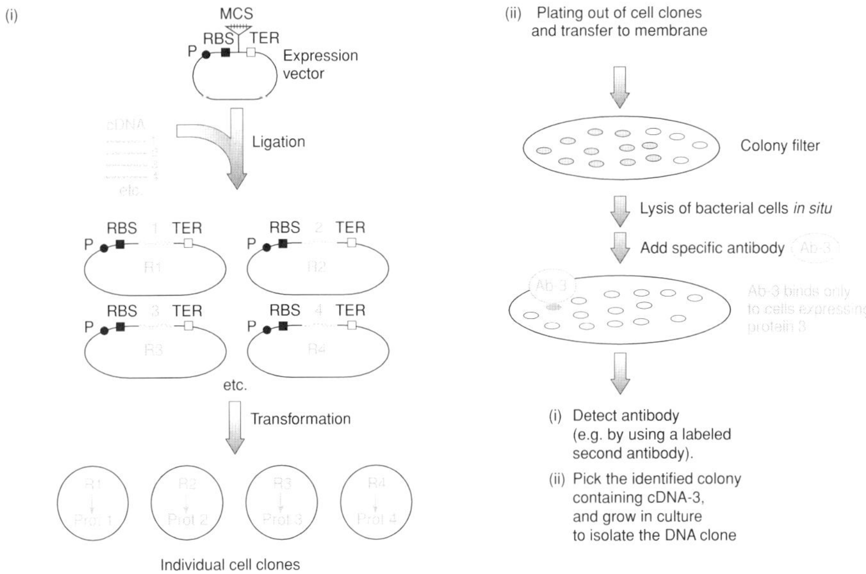

10. **One difference between cloning with M13 vectors and with phagemid vectors is**

A M13 vectors are classical plasmids; phagemids are not
B phagemid vectors require the use of helper plasmids for efficient release of packaged DNA particles from the infected cell; M13 does not
C M13 vectors have two origins of replication (*Col*E1 and f1) while phagemids have only one
D phagemids can be synthesized as a single stranded product; M13 cannot
E two or more of the above are differences between phagemids and M13

Review questions

1. Explain why type II restriction endonucleases are more useful for molecular biologists than type I restriction endonucleases.

2. What are the advantages and disadvantages of cloning into each of the following vectors: pUC-based plasmids, cosmids, lambda phage, bacteriophage P1, and YACs.

Open-ended problems

1. Monoclonal antibodies can be generated to any antigen using well established (and somewhat complex) protocols. You have been given an unlimited supply of a monoclonal antibody that recognizes a molecule (M) on the surface of a lymphocyte that has been activated by a certain cytokine (but is absent otherwise). How could you use that monoclonal antibody to clone the gene for M? Detail the relevant steps. There are many. Also list *relevant* assumptions you are making.

2. *Mushroom-body-miniature-B, mbmB*, is a *Drosophila* mutation that leaves flies with an abnormally small brain and reduced ability to learn using olfactory stimuli. A series of three-point (genetic) mapping crosses has placed the mutation in region 31 on the chromosome 2 in *Drosophila*. A friend who is also studying *Drosophila* has kindly provided you with a 200 bp piece of DNA that she has mapped cytologically (by hybridization to polytene chromosomes) to band 31B, probably very near the *mbm-B* locus! How can you use that 200 bp probe to obtain an *mbmB* cDNA assuming the probe is 30 kb away from the *mbmB* gene?

3. *Figure 4.1* shows a partial sequence of the semaphorin gene found in the vaccinia virus genome. The one-letter code for each amino acid predicted to be encoded by the sequence is indicated below the gene sequence. *Figure 4.2* lists restriction enzymes that would cut the semaphorin gene shown, the recognition sequence for each enzyme and the cleavage sites (^). The cut sites given are with respect to the numbered sequence in *Figure 4.1*.

   (a) Find all the *Bsm*I and *Fok*I recognition sites on *Figure 4.1*. For each restriction site, pencil in the bases of the *non-coding* strand (label 5′ and 3′ ends) and draw a line through the phosphodiester bonds that are broken by each enzyme.
   (b) If *Ava*II's recognition sequence specificity included 5′ G^GCCC 3′ and 5′ G^GGCC 3′, what would *Ava*II be with respect to *Asu*I?
   (c) Which two enzymes could you cut with to get the largest possible piece of the viral semaphorin gene? How big would that restriction fragment be?
   (d) What kind of restriction endonuclease is *Fok*I?
   (e) Do *Mse*I fragments and *Mae*I fragments have compatible ends?
   (f) Do *Sal*I and *Taq*I?
   (g) If you cut the viral semaphorin gene with *Spe*I and *Sal*I and ligate the *Spe*I-*Sal*I fragment into a 4887 bp plasmid, also cut with *Spe*I and *Sal*I (each present once in the multiple cloning site, 51 bases apart), what would the expected size of the plasmid product?
   (h) If you cut the viral semaphorin gene shown in *Figure 4.1* with *Spe*I and *Xba*I and ligate the resulting restriction fragment into a plasmid that has been cut with *Spe*I and *Xba*I (each of which have only one cut site in the plasmid, in the multiple cloning site) what difficulty will you encounter? Would dephosphorylation of the vector eliminate the problem? Why or why not?
   (i) How often should the sequence GTAC occur in a 1080 bp sequence? You may assume that each base is equally abundant, though that is not frequently the case.
   (j) How often does GCGC actually occur? Does the observed number of these sites match the expected number of GCGC sequences? Aside from chance variation from expected frequency, what could explain the discrepancy?

Figure 4.1 Vaccinia virus semaphorin sequence (A39R)

```
GTA TAC GTA CCT GGT AAG ATA TCT TTA TAT TCT ATA ATC AAT GAG ACA TCA CTA   54

TCC GAT AAC GAA TGA AGT CTA GCA CTA GTA TGC CAT TTA CTT AAT ATG GTC GTC  108

TTG GAA GTT TTA TTA TAA GTT AAA ATA TCA TGG TTG TCC AAT TTC CAT CTA ATA  162

TAC TTT GTC GGA TTA TCT ATA GTA CAC GGA ATA ATG ATG GTA TTA TTA CAT GCT  216
                                            M   M   V   L   L   H   A

GTA TAC TCT ATA GTC TTT GTA GAT GTT ATA ATC ATA AAA GTA CAG AGG TAT ATC  270
V   Y   S   I   V   F   V   D   V   I   I   I   K   V   Q   R   Y   I

AAC GAT ATT CTA ACT CTT GAC ATT TTT TAT TTA TTT AAA ATG ATA CCT TTG TTA  324
N   D   I   L   T   L   D   I   F   Y   L   F   K   M   I   P   L   L

TTT ATT TTA TTC TAT TTT GCT AAC GGT ATC GAA TGG CAT AAG TTT GAA ACG AGT  378
F   I   L   F   Y   F   A   N   G   I   E   W   H   K   F   E   T   S

GAA GAA ATA ATT TCT ACT TAC TTA TTA GAC GAC GTA TTA TAC ACG GGT GTT AAT  432
E   E   I   I   S   T   Y   L   L   D   D   V   L   Y   T   G   V   N

GGG GCG GTA TAC ACA TTT TCA AAT AAT AAA CTA AAC AAA ACT GGT TTA ACT AAT  486
G   A   V   Y   T   F   S   N   N   K   L   N   K   T   G   L   T   N

AAT AAT TAT ATA ACA ACA TCT ATA AAA GTA GAG GAT GCG GAT AAG GAT ACA TTA  540
N   N   Y   I   T   T   S   I   K   V   E   D   A   D   K   D   T   L

GTA TGC GGA ACC AAT AAC GGA AAT CCC AAA TGT TGG AAA ATA GAC GGT TCA GAC  594
V   C   G   T   N   N   G   N   P   K   C   W   K   I   D   G   S   D

GAC CCA AAA CAT AGA GGT AGA GGA TAC GCT CCT TAT CAA AAT AGC AAA GTA ACG  648
D   P   K   H   R   G   R   G   Y   A   P   Y   Q   N   S   K   V   T

ATA ATC AGT CAC AAC CCA TGT GTA CTA TCT GAC ATA AAC ATA TCA AAA GAA GGA  702
I   I   S   H   N   G   C   V   L   S   D   I   N   I   S   K   E   G

ATT AAA CGA TGG AGA AGA TTT GAC GGA CCA TGT GGT TAT GAT TTA TAC ACG GCG  756
I   K   R   W   R   R   F   D   G   P   C   G   Y   D   L   Y   T   A

GAT AAC GTA ATT CCA AAA GAT GGT TTA CGA GGA GCA TTC GTC GAT AAA GAT GGT  810
D   N   V   I   P   K   D   G   L   R   G   A   F   V   D   K   D   G

ACT TAT GAC AAA GTT TAC ATT CTT TTC ACT GAT ACT ATC GGC TCA AAG AGA ATT  864
T   Y   D   K   V   Y   I   L   F   T   D   T   I   G   S   K   R   I

GTC AAA ATT CCG TAT ATA GCA CAA ATG TGC CTA AAC GAC GAA GGT GGT CCA TCA  918
V   K   I   P   Y   I   A   Q   M   C   L   N   D   E   G   G   P   S

TCA TTG TCT AGT CAT AGA TGG TCG ACG TTT CTC AAA GTC GAA TTA GAA TGT GAT  972
S   L   S   S   H   R   W   S   T   F   L   K   V   E   L   E   C   D

ATC GAC GGA AGA AGT TAT AGA CAA ATT ATT CAT TCT AGA ACT ATA AAA ACA GAT 1026
I   D   G   R   S   Y   R   Q   I   I   H   S   R   T   I   K   T   D

AAT GAT ACG ATA CTA TAT GTA TTC TTC GAT AGT CCT TAT TCC AAG TCC GCA TTA 1080
N   D   T   I   L   Y   V   F   F   D   S   P   Y   S   K   S   A   L
```

Figure 4.2 Restriction sites in the semaphorin sequence shown in *Figure 4.1*

Enzyme	Recognition	Cut site
*Acc*I	(GT^MKAC)	Def: 3 219 441 941
*Asu*I	(G^GNCC)	Def: 728 911
*Ava*II	(G^GWCC)	Def: 728 911
*Bsm*I	(GAATGCN^)	Def: 791
	(NG^CATTC)	
*Dra*I	(TTT^AAA)	Def: 307
*Eco*RII	(^CCWGG)	Def: 10
*Eco*RV	(GAT^ATC)	Def: 21 973
*Fok*I	(GGATGNNNNNNNNN^)	Def: 533 678
	(^NNNNNNNNNNNNNCATCC)	
*Hha*I	(GCG^C)	Def: none
*Hind*II	(GTY^RAC)	Def: 942
*Mae*I	(C^TAG)	Def: 74 80 927 1008
*Mae*II	(A^CGT)	Def: 6 411 762 944
*Mae*III	(^GTNAC)	Def: 643 656
*Mse*I	(T^TAA)	Def: 96 129 306 429 479 705
*Nla*III	(CATG^)	Def: 141 215 735
*Nla*IV	(GGN^NCC)	Def: 550
*Nsp*I	(RCATG^Y)	Def: 215
*Pfl*MI	(CCANNNN^NTGG)	Def: 573 776
*Rsa*I	(GT^AC)	Def: 9 186 258 672 811
*Sal*I	(G^TCGAC)	Def: 940
*Scr*FI	(CC^NGG)	Def: 12
*Sna*BI	(TAC^GTA)	Def: 7
*Spe*I	(A^CTAGT)	Def: 79
*Taq*I	(T^CGA)	Def: 354 798 941 957 975 1053
*Xba*I	(T^CTAGA)	Def: 1007

4. P-elements have been made that contain the β-galactosidase gene, *lacZ*, preceded by a promoter that requires the presence of a strong enhancer nearby for transcription (and β-galactosidase production) to occur. Such P-elements can be 'mobilized' in *Drosophila* so that they insert in new locations in the genome of the fly bearing the P-element. These mobilized P-elements are often mutagenic when they land in or near a gene. Sometimes they land near strong tissue-specific enhancers. How could you use these P-elements in a search for genes important in embryonic myogenesis (muscle formation)? Please explain in detail.

5. pcDNA is a line of cloning vectors related to pUC19. The pcDNA expression vectors are sold by the biotechnology company, Invitrogen. What restriction sites are in the multiple cloning site of the latest version of pcDNA3? What selectable marker is used? Include a copy of the complete base sequence of vector with the *Bam*HI and *Pst*I sites underlined. (Hint: Invitrogen's homepage can be found on the Internet).

5 Nucleic acid hybridization assays

Multiple choice questions

1. Random-primed DNA labeling involves which of the following?

A Gamma-labeled dATP
B Nondenatured double-stranded DNA for use as template
C Polynucleotide kinase
D Hexamers of random sequence
E Pancreatic DNAse I

2. Theoretical melting temperature of two DNA strands depends on all of the following EXCEPT

A strand length
B GC content
C monovalent cation concentration
D presence of chemical denaturants
E percent CpG methylation

3. All of the following are methods of screening cDNA libraries for genes of interest EXCEPT

A hybridization with a probe corresponding to the gene of interest
B hybridization with an oligonucleotide probe
C hybridization with an enhancer-based probe
D hybridization with a heterologous probe
E complementation screening

4. The restriction map below shows the *Mst*II sites in a portion of the β^A-globin allele. The β^S-globin allele lacks the internal *Mst*II site. Genomic DNA has been isolated from several individuals, cut with *Mst*II, separated by electrophoresis, transferred to nitrocellulose membrane, and probed with the labeled 1.2 kb *Mst*II fragment. Which of the lanes of the blot shown below correspond to the DNA of individual II_1?

(A) 1
(B) 2
(C) 3
(D) 4
(E) 5

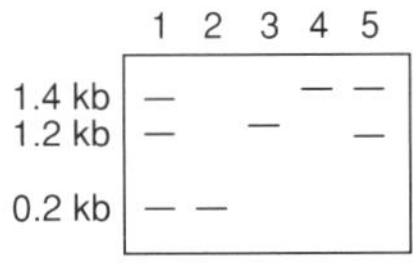

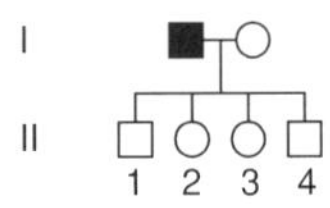

5. Probes generated by run-off transcription from cloned DNA inserts require

A a specialized plasmid vector with promoter upstream of the DNA insert
B DNA polymerase
C dNTPs
D random hexamers
E a labeled deoxyribonucleotide to be incorporated in the probe

6. Which of the following is NOT used in nonisotopic labeling systems?

A Digoxigenin
B ^{32}P
C Biotin
D Fluorescein
E Alkaline phosphatase

7. Colony hybridization involves

A adding a radiolabeled probe to colonies on an agar plate
B adding a radiolabeled probe to a Southern blot
C prior denaturation of both probe and target DNA
D combining colonies of complementary genotypes
E colony blotting and immunoglobulin binding

8. Which of the following temperatures is the closest to the theoretical melting temperature for the probe shown below? You may assume that the hybridization solution will contain 0.1 M monovalent cations (Na^+) and no formamide or urea.

5′ ACGUACGUACGUACGUACGUACGUACGUACGUACGUACGUA 3′

A 25.9°C
B 44.4°C
C 52.9°C
D 82.9°C
E More than 5°C away from any of these values

9. A mutation in the *Crc2A* gene has been implicated in a form of breast cancer. You are probing with two ASOs (allele-specific oligonucleotides) C+ASO directed to the normal *Crc2A* allele and CmASO directed to the mutant *Crc2A* allele. You perform a dot-blot hybridization on a blood sample from a patient with a family history for this particular form of breast cancer. The test results are shown below.

	Patient	+/+	+/m	m/m
Dot blot probed with C+ASO	○	●	●	○
Dot blot probed with CmASO	○	○	●	●

(○ = no hybridization; ● = hybridization)

A A negative result with both ASOs confirms a novel mutation in this region of the Crc2A gene
B A positive result with the C+ASO would have indicated that the patient had two normal copies of the *Crc2A* gene
C A positive result with the C+ASO and a negative result with the CmASO would have indicated one normal and one mutant allele of the *Crc2A* gene
D The patient may have a mutation in the *Crc2A* gene but it is not in the area to which these ASOs hybridize
E This test suggests that the patient has a polymorphism within the area to which these ASOs hybridize

10. RNA from many *Drosophila* developmental stages and tissues has been separated on an RNA gel, blotted to a membrane, and hybridized to a probe made to a *Drosophila* cDNA corresponding to the (imaginary) *mucin* gene. Equal amounts of polyA$^+$ RNA was added to each well of the RNA gel. Which of the statements below is justified based on the Northern blot shown below?

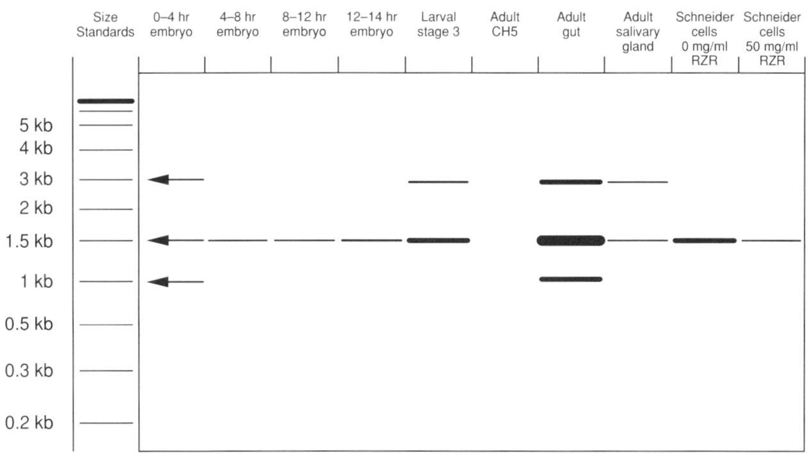

A Mucin expression begins at larval stage 3
B The mucin transcript is most abundant in the adult salivary gland
C The mucin transcript is alternatively spliced
D Mucin expression increases with treatment with the compound RZ32
E None of the above statements are justified

Open-ended problems

1. Design a fully degenerate probe based on the following peptide

   NH_2 I C G T P S Y K N L C R T COOH

   How degenerate would the probe be if no guesses were made as to which codons were most likely to be used at each site? (What is the '-fold degeneracy' of the completely degenerate probe?)

6 PCR, DNA sequencing and *in vitro* mutagenesis

Multiple choice questions

1. Which of the following modifications from standard PCR conditions would most likely destroy the possibility of obtaining the desired product?

A Attaching a digoxigenin molecule at the 5′ end of a primer
B Changing a single base (to a mismatch) at the 5′ end of both primers
C Amplifying from methylated template
D Using a DNA–RNA hybrid as template
E Using ribonucleotides in the reaction

2. How many fold degeneracy would a fully degenerate oligonucleotide primer have that was designed to correspond to the peptide:

$^{+}NH_3$ R H G D G M M COO^{-}?

A 384×
B 256×
C 128×
D 64×
E 32×

3. RT–PCR requires

A a cDNA library
B an RNA dependent DNA polymerase
C light RNAse treatment
D degenerate oligonucleotide primers
E ectopic transacetone

4. How big is the amplification product obtained when PCR is performed with the following primers and template using standard reaction chemistry and conditions?

Primer A: 3′ ACA GTC TTA ATG TAA TCT 5′
Primer B: 5′ TCA TTT ACA CTC CAC ATA 3′

Template:

```
5'  GAG GAT GAT TGT CAG AAT TAC ATT AGA ATA CTG TAC TCT TCA GAA CCG GGG AAA
3'  CTC CTA CTA ACA GTC TTA ATG TAA TCT TAT GAC ATG AGA AGT CTT GGC CCC TTT
    TTA GTT ATT TGC GGG ACC AAT TCG TAC AAA CCC CTC TGT CGG ACG TAC GCA TTT
    AAT CAA TAA ACG CCC TGG TTA AGC ATG TTT GGG GAG ACA GCC TGC ATG CGT AAA
    AAG GTA GGT CCA GTC GGT TTG ACG CAA TCG TCA TTT ACA CTC CAC ATA ATT CTA  3'
    TTC CAT CCA GGT CAG CCA AAC TGC GTT AGC AGT AAA TGT GAG GTG TAT TAA GAT  5'
```

A 147 bp
B 162 bp
C 111 bp
D 27 bp
E No product would form

5. Vu is married to Tuyen who was previously married to Lan (Lan is now deceased). Lan and Tuyen conceived one child together and adopted one child. Vu and Tuyen have also conceived one child together. All members of Tuyen's current family have had DNA fingerprinting done at a single VNTR locus. Unfortunately the laboratory misplaced the sheet that identified which child's DNA sample was which. Which of the following can NOT be determined from the lab's data?

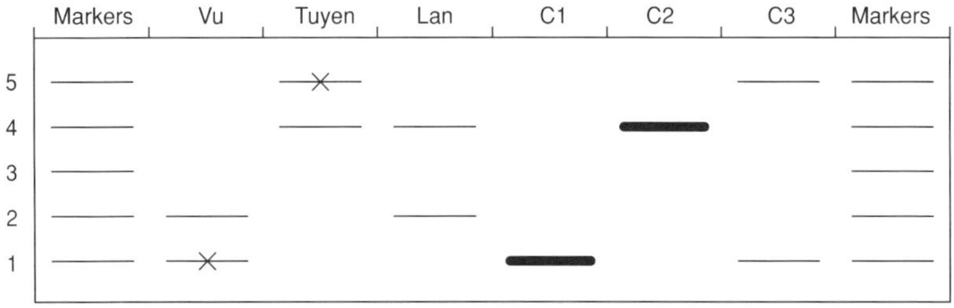

A Child 1 does not have an allele from Tuyen
B Child 2 does not have an allele from Vu
C Child 3 is the best candidate for the child conceived by Vu and Tuyen
D Child 1 is the best candidate for the child adopted by Lan and Tuyen
E Both Child 1 and Child 3 are candidates for the child conceived by Lan and Tuyen

6. The TaqMan procedure allows one to

A assay for the presence or absence of a known point mutation
B amplify microsatellites for rapid determination of genetic fingerprints
C determine a person's sex by PCR
D sequence up to 700 bases using *Taq* polymerase
E none of the above

7. Which of the primer pairs below could be used (along with two primers [not shown] corresponding to the start and end of the open reading frame) to mutate a conserved cysteine to a serine in the section of the cDNA shown below:

```
5′ GAG GAT GAT TGT CAG AAT TAC ATT AGA ATA CTG TAC TCT TCA GAA 3′
3′ CTC CTA CTA ACA GTC TTA ATG TAA TCT TAT GAC ATG AGA AGT CTT 5′
    E   D   D   C   Q   N   Y   I   R   I   L   Y   S   S   E
```

A 5′ GAG GAT GAT TGT CAG AAT TAC ATT AGA ATA 3′ and
3′ CTC CTA CTA ACA GTC TTA ATG TAA TCT TAT 5′

B 5′ GAG GAT GAT TCT CAG AAT TAC ATT AGA ATA 3′ and
3′ CTC CTA CTA AGA GTC TTA ATG TAA TCT TAT 5′

C 5′ GAG GAT TGT TGT CAG AAT TAC ATT AGA ATA 3′ and
3′ CTC CTA ACA ACA GTC TTA ATG TAA TCT TAT 5′

D 5′ GAG GAT TCT TGT CAG AAT TAC ATT AGA ATA 3′ and
3′ CTC CTA AGA ACA GTC TTA ATG TAA TCT TAT 5′

E None of the above

8. Which of the following are features of modern automated sequencing systems?

A Fluorescently labeled sequencing templates
B Capillary electrophoresis separation systems
C Laser-based real-time sequence reading
D Average reads of 10 kilobases
E Two of the above

9. Sanger sequencing reactions require all of the following EXCEPT

A DNA polymerase
B Dideoxynucleotides
C ATP, CTP, GTP and TTP
D A primer
E None of the above (all are required for sequencing)

10. If DNA polymerase I were added to the nucleic acids shown below in addition to the appropriate buffer, magnesium, deoxynucleotides, and label such as [α-^{32}P]dATP, what product would be formed?

5′ AAGAACGCTTGACCCATATCGTCATTCCAGCCGCTTCGATTACAN$_{(300)}$ 3′
and 5′ AAGCGGCTGGAATGACGA 3′

A 5′ AAGCGGCTGGAATGACGA 3′
B 5′ TCGTCATTCCAGCCGCTT 3′
C 5′ AAGCGGCTGGAATGACGATATGGGTCAAGCGTTCTT 3′
D 5′ AAGCGGCTGGAATGACGAAAGAACGCTTGACCCATA 3′
E None of the above

Open-ended problems

1. Describe a specific strategy for replacing, by using PCR-based mutagenesis, the cysteine encoded by bases 544–546 in the vaccinia sequence in *Figure 4.1* with a serine.

7 Organization of the human genome

Multiple choice questions

1. Which of the following is NOT true of the human mitochondrial genome?

A Unlike transcription of nuclear genes, transcription of the mitochondrial DNA starts from promoters in the D-loop region and continues in opposite directions for the two different strands
B The replication of both the H and L strands is unidirectional and starts at specific origins
C The mature mitochondrial RNAs are generated by cleavage of multigenic transcripts
D Recombination in the mitochondrial genome, like that in simple prokaryotes, is mediated by a *recA*-like gene product
E A triple-stranded DNA structure is generated in the D-loop region by synthesizing an additional piece of the H-strand DNA, the 7S DNA

2. Which of the following is almost always true about retrotransposons?

A Retrotransposons have long terminal repeats
B Retrotransposons encode reverse transcriptase
C Retrotransposons don't have duplicated target sequence just outside their border
D Retrotransposons, if provided with *in vitro* packaging extract, would become infection-competent
E None of the above

3. Prior to entering S phase of the cell cycle, human chromosome 1 consists of about

A 250 Mb of DNA
B 100 Mb of DNA
C 25 Mb of DNA
D 2.5 Mb of DNA
E none of the above

4. All of the following approaches have suggested a best estimate of the number of genes in the human genome to be 65 000 to 80 000 EXCEPT

A calculation of 130 genes per band in a 550-band Giemsa stained karyogram
B extrapolation from sequencing of large chromosomal regions
C restriction analysis that puts the total number of CpG islands at about 45 000
D large-scale random sequencing of cDNA clones to generate ESTs which can then be compared to a large set of human genomic coding sequences in sequence databases.
E all of the above have been used to derive the 65 000–80 000 figure

5. Which of the following genes are *often* clustered?

A Genes which encode subunits of the same protein or enzyme
B Genes which encode enzymes in the same metabolic pathway
C Genes which encode isozymes specific for different subcellular compartments
D Genes encoding tissue-specific protein isoforms or isoenzymes
E Genes which encode the same product

6. An organism's genomic DNA is sheared into fragments of approximately 500 bp long, denatured by heating, and then cooled to allow renaturation of the strands (each strand not necessarily with its original H-bonding partner). Based on the C_0t curve below, what fraction of the organism's DNA is *very low copy number*?

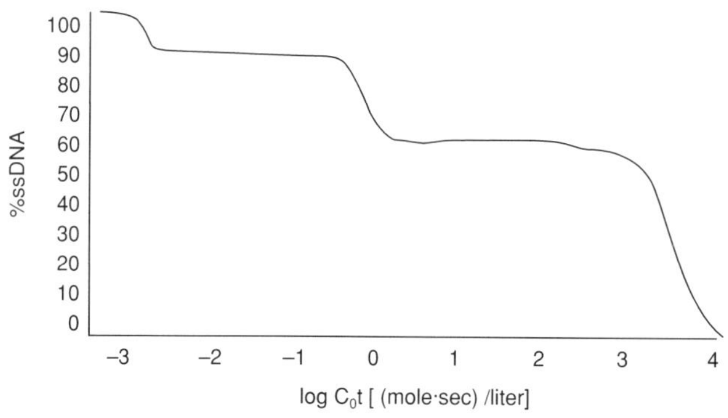

A 1/10
B 1/5
C 3/10
D 1/2
E 3/5

7. Members of DNA sequence families can be identified by which of the following approaches?

A PCR assay
B DNA hybridization assay
C DNA sequencing
D All of the above
E None of the above

8. Satellite DNA

A associates near chromosomes, but doesn't touch them directly.
B is made up of repetitive DNA of different density compared to the bulk of the nuclear DNA
C is clustered exclusively at the far ends of chromosomes
D can undergo secondary amplification to create larger blocks of repeats
E two of the above

9. Which of the following contains microsatellite sequence?

A 5′ ATTGCATTCCATATCATTGCATTCCATATCATGCATTCCATATCAT 3′
B 5′ CATGCTAGGGCAGGACGGGGCAGGATGGGGCAGGAAGGGGCA 3′
C 5′ VNTRCANKLAQEHLMKALTALTMCRMCKGHAGLMPYGGALVNTR 3′
D 5′ TCGATCATCGGATCCAGATCCGCATCGGTTCAAGCATTACGTCCC 3′
E 5′ CAGTCGGCCGATCAGTCAGTCAGTCAGTCAGTCAGTCAGTCAGTC 3′

10. LINE elements differ from *Alu* repeats in all of the following ways EXCEPT

A full-length LINE elements encode reverse transcriptase
B *Alu* elements are more abundant
C *Alu* repeats contain an internal promoter
D LINE elements are located primarily in dark G bands of metaphase chromosomes
E LINE elements have inverted repeats or no repeats at all

Review question

1. What is the difference between a retropseudogene, retrotransposon, and a retrovirus? Give an example of each. What use (evolutionary and modern) might there be for each of these?

Open-ended problems

1. Explain how the following data was obtained, and what it means.

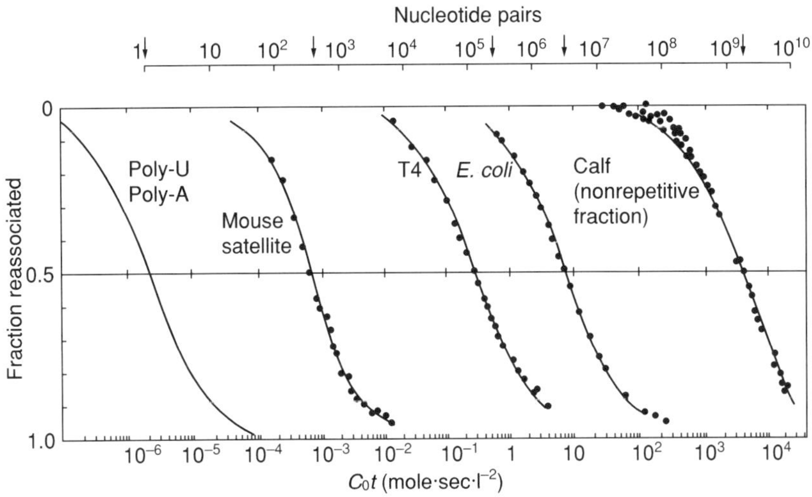

(From Kahl, G. (1995) *Dictionary of Gene Technology*. With permission from Wiley–VCH Verlagsgesellschaft mbH and the author.)

2. Go to GenBank on the Internet and find a gene there in the DEAD box family. Obtain a double-spaced printout of the protein sequence (you don't need to print out the entire gene sequence). Highlight or underline the defining motifs in that protein sequence. From what other species have DEAD box-encoding genes been sequenced?

8 Human gene expression

Multiple choice questions

1. Recognized classes of *cis*-acting transcriptional regulatory elements include all of the following EXCEPT

A silencers
B promoters
C response elements
D transcription factors
E enhancers

2. In the HS-40 α-globin regulatory site shown below, which of the transcription factors listed has the most binding sites? Hint: TF-binding sites are underlined.

```
5′ TCGACCCTCTGGAACCTATCAGGGACCACAGTCAGCCAGGCAAGCACATCTGCCCAAGCC
3′ AGCTGGGAGACCTTGGATAGTCCCTGGTGTCAGTCGGTCCGTTCGTGTAGACGGGTTCGG
AAGGGTGGAGGCATGCAGCTGTGGGGTCTGTGAAAACACTTGAGGGAGCAGATAACTGGGCCAA
TTCCCACCTCCGTACGTCGACACCCCAGACACTTTTGTGAACTCCCTCGTCTATTGACCCGGTT
CCATGACTCAGTGCTTCTGGAGGCCAACAGGACTGCTGAGTCATCCTGTGGGGGTGGAGGTGGA
GGTACTGAGTCACGAAGACCTCCGGTTGTCCTGACGACTCAGTAGGACACCCCCACCTCCACCT
CAAGGGAAGGGGTGAATGGTACTGCTGATTACAACCTCTGGTGCTGCCTCCCCCTCCTGT
GTTCCCTTCCCCACTTACCATGACGACTAATGTTGGAGACCACGACGGAGGGGGAGGACA
TTATCTGAGAGGGAAGGCCATGCCCAAAGTGTTCACAGCCAGGCTTCAG 3′
AATAGACTCTCCCTTCCGGTACGGGTTTCACAAGTGTCGGTCCGAAGTC 5′
```

A GATA-1
B NF-E2
C CACC-1
D OTF-2
E TCF-1

Transcription factors and their binding sequences:

AP2	5′ CCC(A/C)N(C/G)$_3$ 3′
GATA-1	5′ (A/T)GATA(A/G) 3′
NF-E2	5′ TGACTCAG 3′
OTF-2	5′ ATGCAAAT 3′
TCF-1	5′ (C/A)A(C/A)AG 3′

3. Which of the following structural motifs is primarily involved in mediating protein–protein interaction?

A Helix–turn–helix
B Leucine zipper
C Zinc finger
D Dead box
E Recognition helix

4. Transcription factors most often have which of the following pair of domains?

A DNA-binding and activation
B DNA-binding and transcription
C Response and transcription
D Association and transcription
E None of the above

5. Under conditions of iron scarcity in a human cell, the iron response element binding protein is active and

A binds to ferritin H-chain mRNA, protecting it from degradation
B binds to transferrin receptor mRNA, inhibiting its translation
C assumes a hairpin loop configuration
D activates transcription of transferrin receptor
E inhibits translation of an mRNA by binding to its 5′ UTR

6. Which of the following statements best describes the classical view of a gene?

A A transcipt can be edited
B A transcript can be alternately polyadenylated
C A transcript can encode a single polypeptide product
D A transcript can be alternatively spliced
E None of the above

7. Alternative splicing can alter the functional properties of a protein in which of the following ways?

A Making tissue-specific isoforms
B Making both membrane-bound and soluble isoforms
C Localizing protein to different intracellular compartments
D Altering the function of the protein
E All of the above

8. What is the best explanation for why the CpG dinucleotide is under-represented in vertebrate DNA?

A The genome is only about 40% CG
B CpGs tend to mutate to TpGs because of frequency of cytosine methylation
C CpGs are found in the recognition sites of endogenous endonucleases and so are highly regulated
D DNAse I sensitivity
E None of the above

9. Systemic lupus erythematosus (SLE), more commonly known as Lupus is an autoimmune disease that occurs with far greater frequency in women than in men. One of the characteristics of the disease is the production of anti-DNA antibodies in the blood of patients. Which of the hypotheses outlined below could provide a (partial) explanation of the gender-bias in Lupus incidence?

A Lupus is an X-linked dominant trait and would therefore be expected to be more common in females
B The methylation of the Lupus gene is imprinted in females but not in males, thus allowing the disease to appear in females
C Alternative splicing of genes in females, but not in males, allows the Lupus gene to be expressed in blood cells, leading to the disease
D Mosaicism of X-chromosome inactivation in females causes some self-reactive lymphocytes to escape the process of negative selection
E None of the above

10. Transcriptional repression by histone deacetylation is thought to be mediated by

A DNA methylation
B DNA histonylases
C DNA acetylation
D DNAse I
E None of the above

Open-ended problems

1. How would you investigate whether a gene organized as shown below is (a) transcribed from a single promoter and undergoes *cis*-alternative splicing, (b) transcribed from multiple promoters and undergoes *cis*-alternative splicing, (c) transcribed from multiple promoters and undergoes *trans*-alternative splicing, or (d) undergoes DNA rearrangement? These models are shown schematically below.

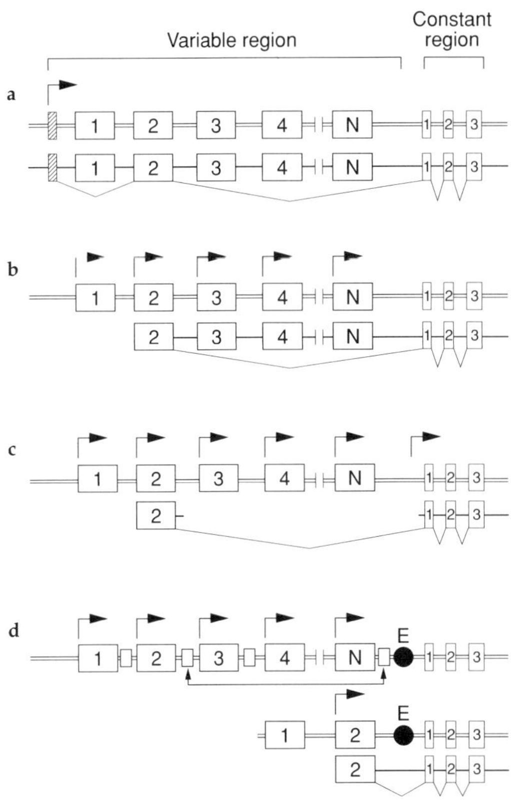

(From Wu, Q. and Maniatis, T. (1999) *Cell* **97**, 779–790, with permission from Cell Press, Cambridge, MA.)

9 Instability of the human genome: mutation and DNA repair

Multiple choice questions

1. A gene that has a significant deletion in its enhancer region but still has a functional promoter would be expected to

A retain tissue-specific expression but lose basal level of expression
B retain regulation of the expression level but lose tissue-specific expression patterns
C retain the basal transcription level defined by the promoter but lose regulation of tissue-specific and stage-specific expression and regulation of expression level
D become incapable of binding RNA polymerase at the enhancer-less promoter
E none of the above

2. Mutations that destroy splice donor, acceptor or branch sites, or that activate a cryptic splice donor or acceptor site, can cause

A failure to remove an intron
B loss of an exon normally retained
C a gene that encodes a truncated protein
D failure of tissue-specific alternative splicing
E all of the above

3. Which of the following shows a transversion and a missense mutation of the sequence:

5′ `ATG AGG TGT CAA`? (The reading frame is indicated by grouping of bases)

A 5′ `ATG AGG TGA CAA`
B 5′ `ATG AGG TAT CAA`
C 5′ `ATG AGG TGC CAA`
D 5′ `ATG AGG TGG CAA`
E Three of the above

4. Triplet expansions are most likely caused by

A unequal crossovers between *Alu* sequences flanking the triplets
B reciprocal translocations
C slipped strand mispairing
D gene conversion
E transposition mediated by retroviruses

5. Monoamine oxidase A is an FAD-containing enzyme encoded in the nucleus but active in the mitochondria. It catalyzes the oxidative deamination of biogenic amines throughout the body. The amines that it degrades include the important neurotransmitters epinephrine and serotonin. It is inhibited by amphetamines, which leads to the 'rush' associated with drugs in the amphetamine family. The entire protein sequence of human monoamine oxidase A is shown below.

```
                    MENQEKASIAGHMFDVVVIGGGISGLSAAKLLTEYGVSVLVLEA
RDRVGGRTYTIRNEHVDYVDVGGAYVGPTQNRILRLSKELGIETYKVNVSERLVQYVK
GKTYPFRGAFPPVWNPIAYLDYNNLWRTIDNMGKEIPTDAPWEAQHADKWDKMTMKEL
IDKICWTKTARRFAYLFVNINVTSEPHEVSALWFLWYVKQCGGTTRIFSVTNGGQERK
FVGGSGQVSERIMDLLGDQVKLNHPVTHVDQSSDNIIIETLNHEHYECKYVINAIPPT
LTAKIHFRPELPAERNQLIQRLPMGAVIKCMMYYKEAFWKKKDYCGCMIIEDEDAPIS
ITLDDTKPDGSLPAIMGFILARKADRLAKLHKEIRKKKICELYAKVLGSQEALHPVHY
EEKNWCEEQYSGGCYTAYFPPGIMTQYGRVIRQPVGRIFFAGTETATKWSGYMEGAVE
AGERAAREVLNGLGKVTEKDIWVQEPESKDVPAVEITHTFWERNLPSVSGLLKIIGFS
TSVTALGFVLYKYKLLPRS
```

The DNA sequence of the most common (wild-type) allele, that corresponds to the amino acids underlined above are as follows

Monoamine oxidase A – partial sequence of the wild-type allele

```
                                                  *
900 - AGACCAGAGCTTCCAGCAGAGAGAAACCAGTTAATTCAGCGGCTTCCAATG - 950
      TCTGGTCTCGAAGGTCGTCTCTCTTTGGTCAATTAAGTCGCCGAAGGTTAC
    - R  P  E  L  P  A  E  R  N  Q  L  I  Q  R  L  P  M
```

Brunner syndrome is an X-linked condition associated with mild retardation, impulsiveness, aggression and sometimes violent behavior. Individuals with the condition typically also get very sick when they eat Chinese food, though the reason for this is not clear. The genetic defect in Brunner syndrome has been mapped to the monoamine oxidase A (MAOA) gene. It appears that at least in one extended family the mutation 936(C→T) underlies the syndrome.

Which of the following is true about the MAOA mutation, 936(C→T)?

A It causes a frame shift that leads to a C-terminally truncated protein
B It changes an amino acid to another amino acid and may therefore destabilize the structure or change the active site of the protein
C It creates a stop codon that would cause the monoamine oxidase A enzyme to be truncated at its C terminus
D It activates a cryptic splice site that causes the deletion of exonic sequence and the production of an internally deleted protein
E There is insufficient data to make any of these conclusions

6. In an effort to design a genetic test for Brunner syndrome in a community where the family bearing the 936(C→T) mutation was discovered, a chip-based array is designed that will allow the detection of the mutation. The oligos below are attached (by their 5′ ends) to chip locations B1, B2, B3, and B4, respectively

```
Oligo 1      5′ CAGAGAGAAACCAGTTAATTC
Oligo 2      5′ CAGAGAGAAACCAGTTAATTT
Oligo 3      5′ CAGAGAGAAACCAGTTAATTG
Oligo 4      5′ CAGAGAGAAACCAGTTAATTA
```

The chip is used to type genomic DNA amplified from three members of a family (pedigree shown below) who have an unexplainably violent son.

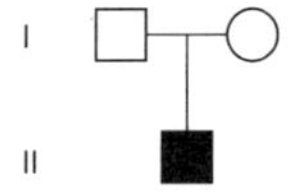

(filled in square = may have Brunner syndrome)

Which of the following patterns best matches the expected SNP-typing results of the individuals shown in the pedigree above?

For each individual (I-1, I-2 and II-2) and for each chip location (B1–B4):

+ means fluorescence detected; – means no fluorescence detected

A I–1 B1+ B2– B3– B4– I–2 B1+ B2+ B3- B4– II–1 B1– B2+ B3– B4–
B I–1 B1– B2+ B3– B4– I–2 B1+ B2+ B3- B4– II–1 B1+ B2– B3– B4–
C I–1 B1– B2– B3+ B4– I–2 B1– B2– B3+ B4+ II–1 B1– B2– B3– B4+
D I–1 B1+ B2+ B3– B4– I–2 B1+ B2– B3– B4– II–1 B1+ B2+ B3– B4–
E I–1 B1+ B2– B3– B4– I–2 B1– B2+ B3– B4– II–1 B1+ B2+ B3– B4–

Open-ended problems

1. Classify the following mutations as completely as you can:

   (a) ATG (the first codon in an ORF) → ATC
   (b) ATG (a codon in the middle of an ORF) → ACG
   (c) CTG (a codon in the middle of an ORF) → TTG
   (d) TGG (a codon in the middle of an ORF) → TGA
   (e) $(CAG)_7$ (a stretch of seven glutamine codons in an ORF) → $(CAG)_{10}$.
   (f) CCG (a codon in the middle of an ORF) → CCAG
   (g) GGC (a codon in the middle of an ORF) → GGT

2. Describe the possible biological bases for the differences in age-related genetic abnormalities in the gametes of males and females.

3. Illustrate in detail how the following genetic change probably came about

7p21 (maternal):	5′	.....ACG TCC CGC ACC AGA TCT TTT.....3′
	3′	.....TGC AGG GCG TGG TCT AGA AAA.....5′
7p21 (paternal)	5′	.....TCG TGC CGC ACC AGT TCT TTG.....3′
	3′	.....AGC ACG GCG TGG TCA AGA AAC.....5′

to

7p21 (maternal):	5′	.....ACG TCC CGC ACC AGA TCT TTT.....3′
	3′	.....TGC AGG GCG TGG TCT AGA AAA.....5′
7p21 (paternal):	5′	.....TCG TCC CGC ACC AGA TCT TTG.....3′
	3′	.....AGC AGG GCG TGG TCT AGA AAC.....5′

4. Describe in detail three mechanisms by which a large tandem repeat can be generated. Illustrate each mechanism you describe.

10 Physical and transcript mapping

Multiple choice questions

1. To make a contig of cosmids corresponding to a large YAC insert one should

A screen a cosmid library with probes made to the insert, purify cosmid clones that hybridize, and order the clones by aligning restriction maps
B label a cosmid, probe for YACs, make a restriction map of YACs and align them end to end
C amplify by PCR one contig at a time using primers designed to amplify the YAC in pieces and then align them by sequencing the ends of each piece
D probe a cDNA library to identify all transcripts in the YAC and align these cDNAs by looking for overlapping ORFs
E none of the above

2. Mapping of ESTs to subchromosomal locations can be done most rapidly by

A using a PCR assay to screen for the presence of a particular EST in a panel of somatic cell hybrids
B recombination mapping in mice using two- or three-point crosses
C isolation of corresponding genomic DNA and use of genomic DNA to probe chromosomes by FISH
D any of the above
E none of the above

3. Which of the following is NOT a method of confirming that a candidate gene is likely to be the locus of gene with a disease allele?

A Restore a normal phenotype to defective cells *in vitro* by transformation of defective cells with the candidate gene
B Find mutations in the gene in people with the disease that are not also found in people without the disease
C Demonstrate the expression of the gene in people with the disease, and compare it to the expression of a second, reference gene that is expressed in normal people
D Produce a model of the disease by deleting the homologous gene in mice
E Identify chromosomal breaks that disrupt the gene in a subset of people with the disease

4. To obtain a contig of cosmids corresponding to a YAC insert one should

A screen a cosmid library with probes made to the insert, purify cosmid clones that hybridize, and order the clones by aligning restriction maps or STS patterns
B label a cosmid, probe for YACs, make a restriction map of YACs and align them end to end
C amplify one contig at a time using PCR with primers designed to amplify the YAC in pieces; then align the contigs by sequencing the ends of each piece
D probe a cDNA library to identify all transcripts in the YAC and align these cDNAs by looking for overlapping ORFs
E none of the above

5. Positional cloning requires

A a known gene product
B a functional assay
C a general idea of molecular pathogenesis
D a mouse model
E a chromosomal location

6. Mapping of ESTs to subchromosomal locations can be done most rapidly by

A using a PCR assay to screen for the presence of a particular EST in a panel of somatic cell hybrids
B recombination mapping in mice using two- or three-point crosses
C isolation of corresponding genomic DNA and use of genomic DNA to probe chromosomes by FISH
D any of the above
E none of the above

7. Most of the following are methods of demonstrating that a candidate gene is likely to be the cause of a particular disease when the gene is mutated. Which is NOT?

A Expression of the gene in people with the disease
B Finding mutations in the gene in people with the disease that are not also found in people without the disease
C Restoration of normal phenotype *in vitro* by transformation with the candidate gene
D Production of a mouse model of the disease by deleting the homologous gene in mice
E Identification of chromosomal breaks in the gene in a subset of people with the disease

8. Pulsed field gel electrophoresis (PFGE) has a major advantage over conventional electrophoresis. The advantage is

A in PFGE two modes of separation are involved: a pulse-chase and an electric field
B with PFGE very small DNA molecules can be separated with excellent resolution
C alternating electric fields of orthogonal orientation can be used to separate very large pieces of DNA in PFGE
D all of the above
E none of the above

Review question

1. Distinguish between an EST, an STS, and a microsatellite. About how many of each have been mapped in the human genome to date? What are the ESTs, STSs, and microsatellites closest to 3p21.3 in the human genome? Why would it be useful to know this?

Open-ended problems

1. A cDNA probe you have will hybridize to hybrid cell lines HHW1405, HHW1499, HHW1064, HHW1118, HHW1138, and HHW105, but NOT to HHW1124, HHW1600, HHW1113, HHW1421, and HHW1452. Where is the gene corresponding to that cDNA found in the human genome? See *HMG2, Figure 10.2* (below) for more information on each of these hybrid cell lines.

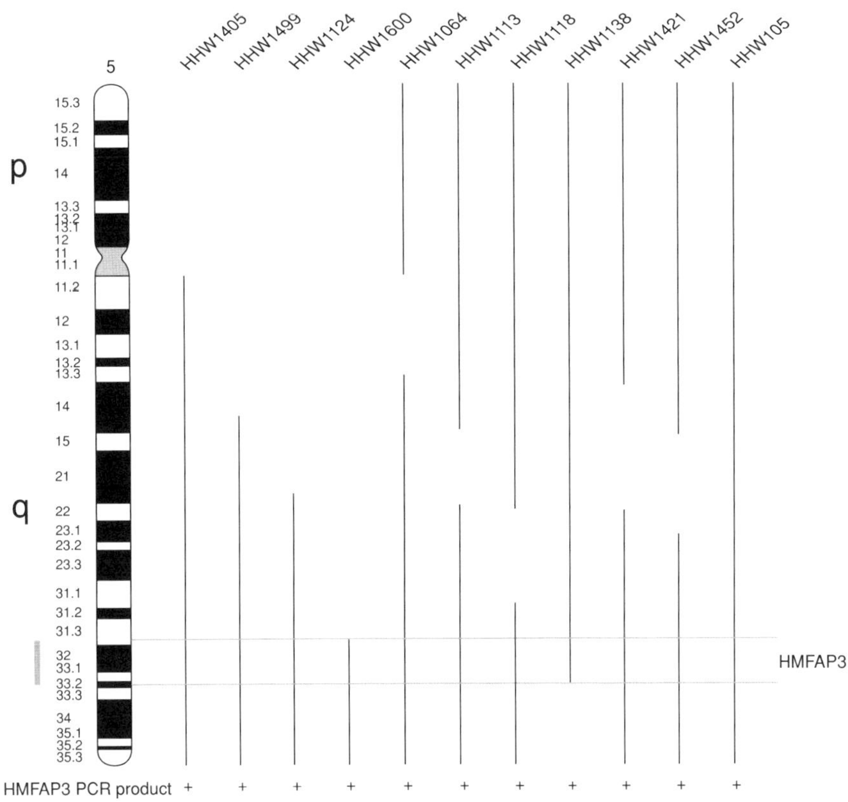

2. You are interested in the gene(s) that might be found in the 100 kb *Drosophila* genomic DNA fragment cloned into DS07441, a P1 vector you have just received from the Berkeley Drosophila Genome Project. Describe how you would identify the genes encoded by the *Drosophila* fragment. Hint: the *Drosophila* genome is (almost) entirely sequenced, and your P1 has already been mapped to polytene chromosomes.

11 Genetic mapping of mendelian characters

Multiple choice questions

1. **Which of the pedigrees below shows an *informative* meiosis for marker A? Shading indicates the individuals with the disease to be mapped.**

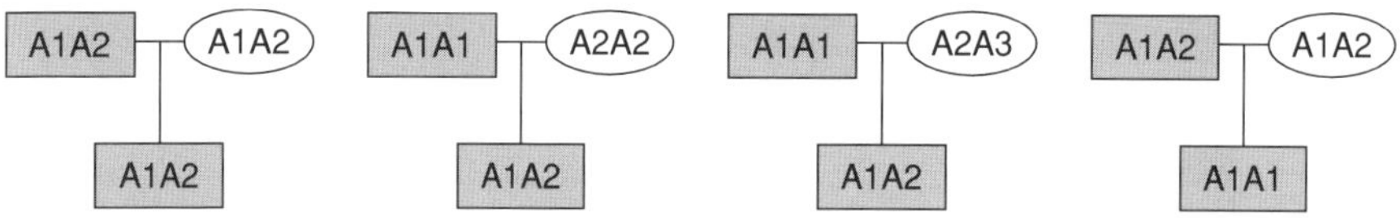

A B C D

E None of the above

2. **Careful analysis of allele frequencies for two neighboring loci, A and B, reveal the following allele frequencies: p(A1) = 0.31, p(A2) = 0.69 and p(B1) = 0.78, p(B2) = 0.11 and p(B3)= 0.11. Which of the two loci would be the better marker for helping map a disease in the vicinity?**

A A is much better
B B is much better
C Both markers are nearly identical in usefulness
D Neither marker is polymorphic
E There is not enough data given to determine

3. **What can you determine from the lod score curve shown below?**

A There is evidence for linkage with nonrecombinants
B There is evidence for linkage with the most likely recombinant fraction of 1/5
C Linkage is excluded for this marker
D Linkage at 0.2 map units
E Evidence for linkage is inconclusive

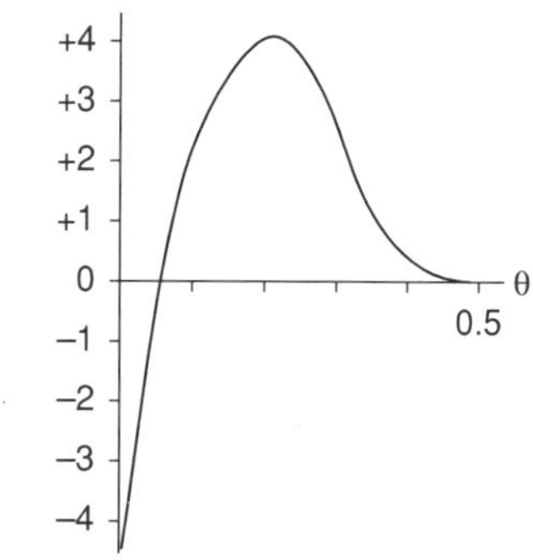

4. For loci 25 map units apart, one would expect a recombinant fraction closest to which of the following? Hint: you should take into account the probability of double and even quadruple crossovers occurring in such a large interval.

A 0.125
B 0.23
C 0.25
D 0.27
E None of the above

5. Which of the following is not a feature of an ideal genetic map?

A DNA samples of reference families widely available
B High confidence in marker order
C High marker density
D Easily typed markers such as VNTRs
E High homozygosity of each marker, preferably near 1.0

6. Allelic association differs from genetic linkage in that

A allelic association means that the alleles on two homologous chromosomes are the same
B linkage of loci A and B implies the association of alleles A1 and B1, A2 and B2, etc.
C most members of the same species will show similar patterns of linkage but, for polymorphic loci, won't show identical patterns of allelic association
D unrelated individuals with the same disease won't tend to show similar linkage patterns but will tend to share the same alleles at loci near the locus with the disease-causing allele
E none of the above

Review question

1. What is the difference in the interpretation of a Z value of 0 and a Z value of –3? What factors could influence the lod calculation to have such values? Distinguish each of the following: theta, map distance and recombination fraction?

Open-ended problems

1. You have a choice of two markers that you could use in the final stage of a genetic mapping project in humans. Both marker loci have been extensively characterized so that you know the allele diversity and frequencies. Marker D2R3001 has alleles A1, A2, A3, and A4 that occur at the relative frequencies of 0.1, 0.1, 0.1 and 0.7. Marker D2R3002 has alleles B1, B2 and B3 at the relative frequencies of 0.3, 0.3, and 0.4. Which of these markers would be more informative? Support your choice quantitatively.

2. In search of a polymorphic marker that is linked to a gene that predisposes mice to obesity, you type 500 obese and nonobese mice for several microsatellite marker alleles.

   For three of the markers' alleles you get the following data:

Allele	Recombinant (*n*)	Nonrecombinant (*n*)
X1-1	10	490
X2-6	42	458
X3-7	150	350

   Calculate and graph the lod score, Z, for the following values of theta for each of the above alleles: 0, 0.1, 0.2, 0.3, 0.4, 0.5. Is there support for linkage? If so, at what distance from the marker loci? If not, what could you do to continue the search?

12 Genetic mapping of complex characters

Multiple choice questions

1. The number of meiotic steps that have occurred since individuals last shared a common ancestor is ________ related to the median length of shared DNA segment they will typically share.

A linearly
B exponentially
C inversely
D not
E none of the above

2. Which of the following is the best measure of linkage disequilibrium for alleles at locus AB403 and locus AB409? The relative frequency of AB403 allele 1 on chromosomes carrying AB409 allele 1 is 0.90. The relative frequency of AB403 allele 1 on chromosomes carrying AB409 allele 2 is 0.10.

A 0.09
B 0.10
C 0.80
D 0.97
E 1.00

3. If n questions are asked in an association study (e.g. if allelic association is checked with 10 marker loci in 10 people with a disease, there are 100 questions asked), what is the appropriate p value to use to measure the threshold of significance? You may assume that $p = 0.05$ would normally have sufficed.

A 0.001
B 0.005
C 0.01
D 0.05
E None of the above

4. Given the particular strengths of association tests and linkage mapping, a natural study design to localize a genetic susceptibility factor is to

A start with a genomewide screen by linkage analysis and then, once an initial localization has been achieved, to narrow down the candidate region by linkage disequilibrium mapping
B perform both linkage analysis and association studies simultaneously to see which one works best; this would be most cost-efficient, too
C start with a genomewide association study and then, once an initial association is achieved, narrow down the candidate region by linkage analysis
D neither of these are well suited for the study of genetic susceptibility factors. It would be better to stick to physical mapping
E none of the above

5. In the four pedigrees below, which, if any, show allele A1 as IBD in the sib pairs?

A B C D

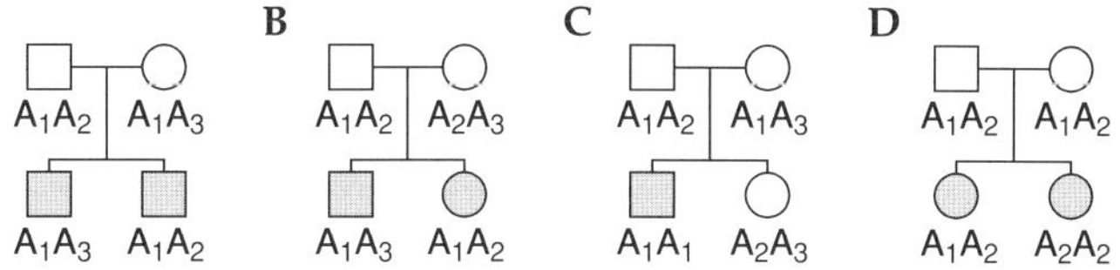

E None of the above

Review questions

1. What are some of the differences between the transmission of disequilibrium test and classical disease-marker association studies?

2. What are a few of the historical reasons why the results of disease association studies have often failed to be reproducible by other investigators?

Open-ended problems

1. The following questions address the issue of sample size (number of parents of affected individuals required by the study) needed to find linkage between a marker locus and the locus of a susceptibility allele using the transmission disequilibrium test.

   (a) What p value is required for establishing 'significant linkage' between a marker allele and a susceptibility allele?

   (b) What is the relative risk conferred by one copy of the susceptibility allele given the following allele frequency data?

      proportion of affected individuals with the allele = 0.7
      proportion of affected individuals without the allele = 0.3
      frequency of the allele in unaffected (control) individuals = 0.1
      frequency of absence of allele in unaffected (control) individuals = 0.9

   (c) What is the probability that a parent of an affected individual will be heterozygous for the susceptibility allele in part (b)?

   (d) What is the probability that a heterozygote parent (identified in (c)) will transmit the high-risk allele to an affected child?

   (e) What sample size would be needed for 80% power (i.e. being able to reject the null hypothesis when it is false) to detect significant association in a genome-wide search using the transmission disequilibrium test given the following data and assumptions?

   (f) What are the assumptions of the model you used to determine the sample size needed?

13 Genome projects

Multiple choice questions

1. Which of the following is <u>NOT</u> a 15-year goal of the US Human Genome Project?

A Construction of a high resolution genetic map of the human genome
B Production of a variety of physical maps of all human chromosomes
C Determination of the complete sequence of human DNA and of the DNA of selected model organisms
D Development of capabilities for collecting, storing, distributing, and analyzing the data produced
E Determination of the function of all human genes by two hybrid analysis

2. With respect to the partial physical maps of chromosome 21 below, which of the following statements is correct?

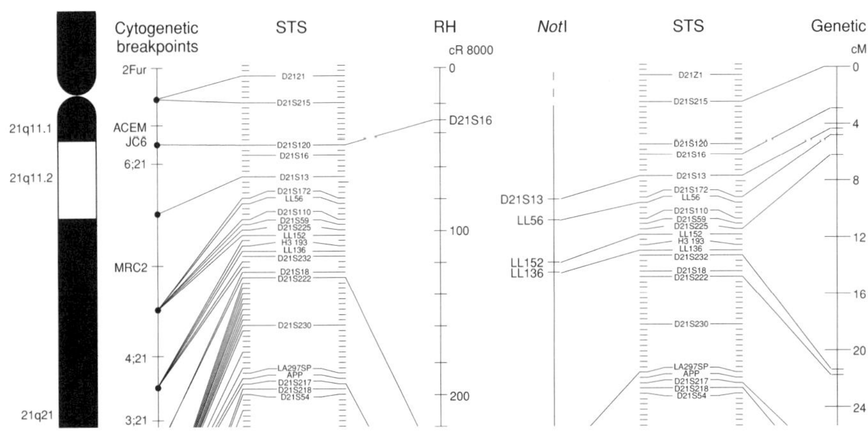

A D21S16 is less than 40 cR_{8000} from the break point in chromosome 2Fur
B Band 21q11.2 is centromeric to 21q11.1
C D21S215 is more than 4 cM from D21S16
D The break in chromosome 21 in 2Fur is telomeric to the break in 6;21 translocation shown
E D21S13 is present in somatic cell hybrids containing only one human chromosome: chromosome 21 containing part of chromosome 6 (i.e. 21t6) whose breakpoint is shown on the map of cytogenetic breakpoints

3. Information from the genome project could allow one to study the function of a newly identified gene by

A overexpression of the gene
B generating a knockout mutation in the homologous gene in mice
C rescuing an organism defective in the gene by transforming with the gene
D all of the above
E none of the above

4. Which of the following represents an expressed sequence tag on chromosome 3?

A D3S213
B E3S213
C D3E213
D D3S213E
E None of the above

5. Which database would be most useful for obtaining a quick, referenced summary of the literature on a human disease?

A CEPH
B GDB
C OMIM
D MGD
E None of the above

Review questions

1. Distinguish between clone-based and disease-based genomics. How do they overlap? What are the advantages of each?

2. What are the compelling reasons for sequencing the genomes of each of the following: *Haemophilus influenzae, Mycobacterium genitalium, Methanococcus jannaschii, E. coli, Saccharomyces cerevisiae, Caenorhabditis elegans, Drosophila melanogaster, Mus musculus, Fugu rubripes rubripes* and *Zea mays*.

Open-ended problem

1. Tumor necrosis factor-α (TNF-α) is a cytotoxic monokine produced by macrophages stimulated with endotoxin (a compound derived from bacterial cells). It provides an immediately active means of defence against infection but is fatal when produced in large quantities. Mutations in the TNF promoter that allow the abnormal binding of the transcription factor, OCT-1, are found in about 5% of Africans and have been linked with an increased susceptibility to cerebral malaria. How could a gene chip (DNA microarray) designed to detect SNPs type a person for the TNF_{-376A} allele? If hybridization is involved, what precise oligo sequence(s) would be used?

14 Our place in the tree of life

Multiple choice questions

1. The mitochondrial genome and the nuclear genome are similar in that

A both have genes that are interrupted by introns
B both have less than 10% of their DNA coding for RNA or proteins
C both encode their own tRNA molecules
D both replicate using bidirectional replication forks
E none of the above

2. Concerted evolution of gene paralogs is thought to involve which of the following?

A Sequence homogenization
B Gene conversion
C Homologous recombination
D Unequal crossing over
E All of the above

3. Which of the following is NOT a 'footprint' of evolution?

A Different chromosome arms bearing similar genes in a similar order
B Processed pseudogenes
C Similar banding patterns on gorilla and human chromosomes
D Ancient proteins from which the phosphate–sugar backbone of DNA is derived
E All of the above

4. The evidence for the late (rather than early) origin of introns includes

A the lack of correspondence between exons and structural and function domains in gene products
B the general lack of conservation of exon sequences
C the fact that almost all eukaryotic genes have introns
D in the postulated ancient RNA world introns would diversify more quickly than in a DNA–RNA–protein world
E three of the above

5. The Y chromosome is essential for the development of human males because

A many genes necessary for viability are on the Y chromosome
B genes for testis formation and spermatogenesis are on the Y chromosome
C the pseudoautosomal region contains genes that are highly dosage sensitive
D three of the above
E none of the above

Review questions

1. Distinguish between the serial endosymbiont hypothesis and the syntrophy hypothesis. What is evidence for each? What evidence needs to be obtained to refine them?

2. What is the history of the human X chromosome beginning with the emergence of mammals? Where is it going and why?

Open-ended problem

1. While *Caenorhabditis elegans* has four semaphorin genes among its 20 000 genes, humans have around 20 in a genome of about 100 000 genes. What mechanisms could explain this discrepancy? What evidence is there for each mechanism?

15 Identifying human disease genes

Multiple choice questions

1. The primary difference between functional cloning and positional cloning is

A in functional cloning you start with a family or association study
B in positional cloning the gene you are looking for has a position; genes isolated by functional cloning are frequently transposed or translocated
C functional cloning is often helped by candidate analysis, which is not useful in positional cloning
D in positional cloning pathogenesis or biochemical functional data are required
E in functional cloning one starts with a DNA or protein; in positional cloning one starts with an approximate subchromosomal location

2. A peptide composed mostly of which of the following groups of amino acids would be most helpful for guiding the design of a low-degeneracy oligonucleotide probe for use in screening for orthologous sequences?

A valine, arginine, glycine
B cysteine, tryptophan, methionine
C serine, threonine, alanine
D isoleucine, proline, serine
E none of the above

3. In the family pedigree of Pendred syndrome (congenital deafness and thyroid goiter) shown below, cross-over analysis provides evidence against which chromosomal region being the possible site for the Pendred syndrome locus?

A D7S1530 → D7S501
B D7S501 → D7S2456
C D7S3073 → D7S3074
D D7S2456 → D7S2425
E None of the above

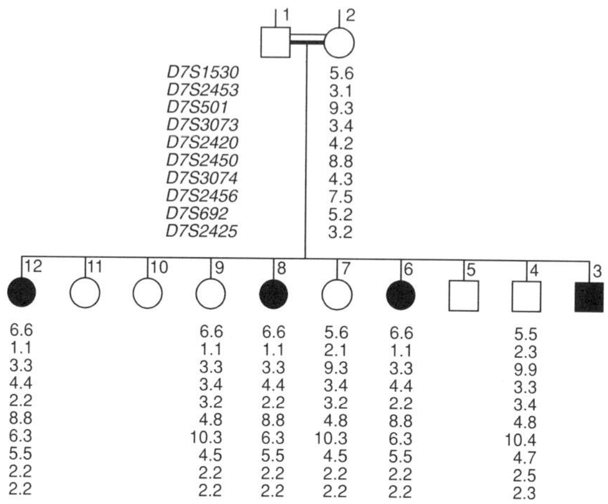

4. Given the somatic cell hybridization results shown below, what is the sub-chromosomal locus of the SIX3 gene?

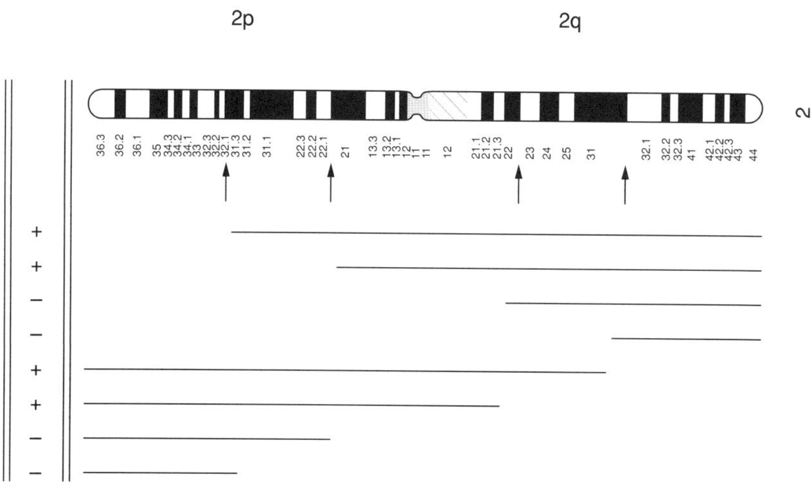

A Between 2pter and 2p32.1
B Between 2p31.3 and 2p22.1
C Between 2p21 and 2q22
D Between 2q23 and 2q31
E Between 2q32.1 and 2qter

5. Individuals with primary thrimethylaminuria have a body odor reminiscent of rotting fish, due to excessive excretion of trimethylamine (TMA). The disorder, also known as fish-odor syndrome, is inherited recessively. Recently a candidate gene was discovered and named flavin-containing mono-oxygenase 3 (*FMO3*). In search of mutations, the *FMO3* gene was sequenced in an individual with fish odor syndrome, his unaffected father, and an unrelated individual (who you may assume is homozygously normal at the *FMO3* locus). Genomic sequence from exon 4 for each of the three is shown on the traditional sequencing gels below.

The wild-type sequence is:

```
5′ GTG TAT CCC AAC CTA CCA AAA AAG TCC TTT 3′
```

NOTE: the three gels are not labeled!

What peptide changes are observed in the man with fish odor syndrome compared to the wild-type?

A Pro → Tyr and Lys → Glu
B Glu → Lys and Leu → Pro
C Pro → Leu and Glu → Lys
D Lys → Glu and Pro → Leu
E None of the above

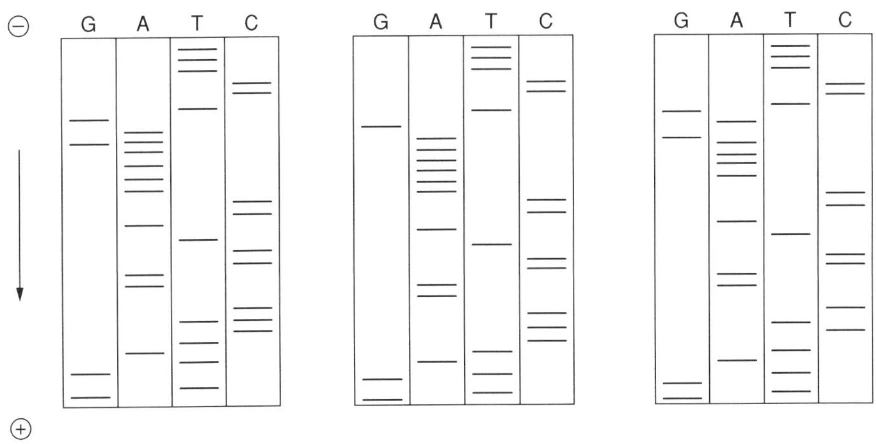

Review questions

1. Define the difference between positional cloning and functional cloning. What are the advantages of each? Give examples of genes that have been cloned by each method.

2. What are the ways in which one can confirm that a candidate disease gene is actually a disease gene?

Open-ended problem

1. After becoming fabulously wealthy you become obsessed with the desire to discover the gene(s) responsible for the innate love of water (love of swimming, absence of fear when near water, tendency to vacation near water, etc.). You plan to fund scientists to pursue this question for you. What precisely would these scientists do?

16 Molecular pathology

Multiple choice questions

The following *two* questions refer to the partial gene sequence conceptually translated in three frames shown on the following page. The start of transcription is at base number 1 which is not shown. The start of translation is at base number 105. In all cases, the last digit of the number is directly above the base to which the number refers.

Sequence data for questions 1 and 2

```
                                051                                 078
TGC ACT TTA ATC CAA TTG AGG TTG GTG GGG TCG CAT TCG CCC CCC GGT CGA CCA
C   T   L   I   Q   L   R   L   V   G   S   H   S   P   P   G   R   P
 A   L   .   S   N   .   G   W   W   G   R   I   R   P   P   V   D   H
  H   F   N   P   I   E   V   G   G   V   A   F   A   P   R   S   T   T

                                105                                 132
CCC CTC TCG CTC GCA CCG CCC TCG CAA TGT CTC TTC TAC AGC TAT CGC CGC TCC
P   L   S   L   A   P   P   S   Q   C   L   F   Y   S   Y   R   R   S
 P   S   R   S   H   R   P   R   N   V   S   S   T   A   I   A   A   P
  P   L   A   R   T   A   L   A   M   S   L   L   Q   L   S   P   L   L

                                159                                 186
TCG CAC TCC TGC TAC TCC TCT GCA GTA GTG TGA GCG AGA CGG CTG CGG ACT ACG
S   H   S   C   Y   S   S   A   V   V   .   A   R   R   L   R   T   T
 R   T   P   A   T   P   L   Q   .   C   E   R   D   G   C   G   L   R
  A   L   L   L   L   L   C   S   S   V   S   E   T   A   A   D   Y   E

                                213                                 240
AGA ACA CCT GGA ACT TCT ACT ACG AGC GTC CCT GTT GCA CTG GAA ACG ATC AGG
R   T   P   G   T   S   T   T   S   V   P   V   A   L   E   T   I   R
 E   H   L   E   L   L   L   R   A   S   L   L   H   W   K   R   S   G
  N   T   W   N   F   Y   Y   E   R   P   C   C   T   G   N   D   Q   G

                                267                                 294
GGA ACA ACA ATT ACG GAA AAC ACG GCG CAG ATC ATG TGC GGG AGT TCA ACT GCG
G   T   T   I   T   E   N   T   A   Q   I   M   C   G   S   S   T   A
 E   Q   Q   L   R   K   T   R   R   R   S   C   A   G   V   Q   L   R
  N   N   N   Y   G   K   H   G   A   D   H   V   R   E   F   N   C   G

                                321                                 348
GCA AGC TGT ACT ATC GTA CAT TCC ATA TGA ACG AAG ATC GAG ATA CGC TCT ATG
A   S   C   T   I   V   H   S   I   .   T   K   I   E   I   R   S   M
 Q   A   V   L   S   Y   I   P   Y   E   R   R   S   R   Y   A   L   C
  K   L   Y   Y   R   T   F   H   M   N   E   D   R   D   T   L   Y   V

                                375                                 402
TGG GAG CCA TGG ATC GCG TAT TCC GTG TGA ACC TGC AGA ATA TCT CCT CAT CCA
W   E   P   W   I   A   Y   S   V   .   T   C   R   I   S   P   H   P
 G   S   H   G   S   R   I   P   C   E   P   A   E   Y   L   L   I   Q
  G   A   M   D   R   V   F   R   V   N   L   Q   N   I   S   S   S   N
```

1. Which of the following would cause a frameshift that would truncate the protein encoded by the gene?

A 51G→A
B 279C→A
C 131-132insC
D 148-150delTCC
E None of the above

2. What would be the most likely consequence of 105A→T?

A Glutamine replaced by a histidine
B Alanine replaced by a threonine
C Methionine replaced by leucine
D An N-terminally truncated protein or no protein made
E None of the above

3. For collagen genes, a missense mutation can be more severe than a nonsense mutation because

A some collagen missense mutations will lead to aberrantly packed protein
B the nonsense mutation will lead to a complete loss of collagen protein in the body while the missense mutation will just reduce the amount
C as a rule, amorphic mutations lead to less severe phenotypes than hypomorphic ones
D nonsense collagen mutations can be compensated for almost entirely by increased transcription
E none of the above

4. All of the following are mechanisms of gain-of-function mutations EXCEPT

A acquire new substrate (e.g. Pittsburg allele)
B overexpression (e.g. PMP22)
C receptor permanently on (e.g. GNAS1)
D deletion of promoter sequence (e.g. Hb-alpha)
E all of the above are known gain-of-function mutations

5. Of the following kinds of genetic mutations, which category is most likely to contain dominant-negative mutations?

A Amorph
B Hypomorph
C Hypermorph
D Antimorph
E Neomorph

6. Thallasemia, a common cause of anemia, can arise from

A the loss of a single alpha-globin allele
B instability of fetal hemoglobin gene products
C a loss-of-function mutation in thalloglobin
D the possession of one or two functional alpha-globin alleles
E none of the above

Review questions

1. What is the difference between loss-of-function and gain-of-function mutations? List five ways each can come about. For each of the following say whether it is due to a gain-of-function or loss-of-function mutation: achondroplasia, cystic fibrosis, Huntington's disease, lactose tolerance.

2. What is the difference between epigenetic and dominant with respect to a locus or allele?

3. What is the difference between atavism and anticipation? Hint: you may need to look up atavism in a standard dictionary.

4. What are the ways of classifying mutations based on their effects on the phenotype?

5. What is a compound heterozygote?

Open-ended problems

1. Williams syndrome is due to a deletion of 1.5–2 Mb of DNA on chromosome 7p11.23. The region deleted is shown below together with genes known to be deleted in most cases of the syndrome.

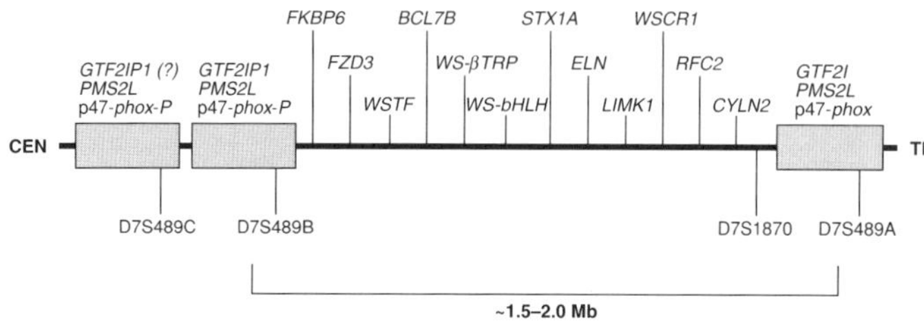

(From DeSilva *et al.* (1999) *Genome Research* **9**(5), 428–436. With permission from Cold Spring Harbor Laboratory Press and the authors.)

(a) What is the best term to describe a genetic disease such as Williams syndrome?
(b) Make a complete list of components of the phenotype of Williams syndrome and a complete list of genes (by their full name), showing wherever possible the link between specific genes and specific features of the syndrome.
(c) Hypothesize why this particular deletion is so common (1 in 20 000 to 1 in 50 000 births). Hint: recall some of the mechanisms of genomic evolution.

2. Comment as fully as you can on the nature of each of the following mutations based on the data provided.

(a) When homozygous mutation 1a has a phenotype that is half as severe as a homozygous null mutation of the same locus.
(b) Mutation 1b when heterozygous with the wild-type allele yields a more severe phenotype than an organism heterozygous for a null mutation.
(c) Mutation 1c is on the X chromosome; men who have it tend to have a wide variety of problems that cannot be explained by simple pleiotropy.
(d) Homozygotes for mutation 1d1 have congenital blindness. Homozygotes for mutation 1d2 are also congenitally blind. All the children of a man and a woman of these two genotypes, however, can see.
(e) Mutation 1e tends to be observed as a somatic mutation. A protein that is normally expressed at low levels is found at constitutively high levels and is associated with unrestrained growth of the cell the mutation is found in.
(f) When mutation 1f is maternally inherited, the affliction is characterized by inappropriate laughter and jerky movements. When the same mutation, 1f, is paternally inherited, the affliction is characterized by weakness and obesity.
(g) For mutation 1g there is the normal amount of mRNA produced but a reduced amount of a protein that is smaller than usual.
(h) Tumor cells with a single copy of mutation 1h contain a transmembrane protein that now possesses a cytoplasmic kinase domain that is not part of the normal transmembrane protein.
(i) Though the predicted amino acid change in the protein encoded by mutation 1i is conservative and doesn't appear to correspond to a crucial position in the protein's active site or structural motifs, the mRNA is unusually large and the protein is unusually small (or absent).

17 Genetic testing in individuals and populations

Multiple choice questions

1. In reference to the pedigree (where the filled in square represents a boy with X-linked muscular dystrophy) and PCR typing data using the intragenic marker *STR45*, shown below, in which individual in this pedigree did the mutation first appear?

A I1
B I2
C II1
D II2
E III1

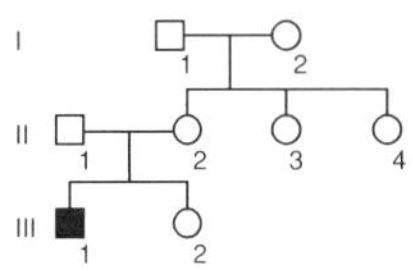

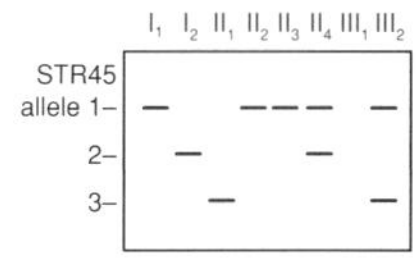

2. Williams syndrome is a complex disease characterized by mental retardation (average IQ 60), a rare heart defect (supravalvular aortic stenosis; SVAS), joint abnormalities, premature aging of the skin, a hoarse voice, elfin facial features, hypercalcemia, extreme friendliness, tremendous curiosity, extraordinary language skills and vocabularies, low spatial reasoning ability, and a near perfect memory for anything they hear. It is thought to be a contiguous gene deletion syndrome.

Recently the gene responsible for SVAS was found on chromosome 7. The gene encodes elastin, a structural protein important for the development of elastic fibers in arteries, lungs, and skin. Persons lacking one copy of the elastin gene develop SVAS but have normal intelligence, personalities, and spatial abilities. They do not, however, have Williams syndrome.

How would you go about identifying the gene whose loss leads to the spatial reasoning deficit in Williams syndrome?

A Use an expression cloning strategy with an antibody made to the region of the brain known to be involved in spatial reasoning
B Use a subtractive hybridization approach to identify cDNAs expressed in Williams' syndrome patients at a higher level than in normal individuals
C Use the elastin gene as a starting point in a positional cloning procedure to identify nearby genes that can be knocked out in mice to be tested for spatial reasoning
D Use an association study to identify marker alleles associated with the spatial reasoning deficit which can then be used (if the LOD score is > 3) to identify the underlying gene
E None of the above

3. When screening for mutations in a human disease gene without first sequencing exons from normal and affected individuals, all of the following can be used effectively EXCEPT

A single-strand conformational polymorphism analysis (SSCP)
B chemical cleavage of mismatches (CCM)
C protein truncation test (PTT)
D enzymatic cleavage of mismatches
E amplification refractory mutation system (ARMS).

4. Situations that make the identification of mutations (and the interpretation of mutation screening) more difficult include all of the following EXCEPT

A mutational heterogeneity
B mutations that are not unambiguously pathogenic
C unsuspected locus heterogeneity
D mutations in a large gene that are hard to find
E none of the above

5. ARMS can be used to test whether a person carries a specific mutation in given gene. Analyze the test results of individuals X1, X2 and X3 (gels shown below) with respect to their genotype at the B locus (BB = normal, Bb = carrier, bb = diseased). The b allele is due to the following mutation: 1394(G→C).

A X1=BB, X2=Bb, X3=bb
B X1=BB, X2=bb, X3=Bb
C X1=Bb, X2=BB, X3=bb
D X1=Bb, X2=bb, X3=BB
E Cannot be determined from this data

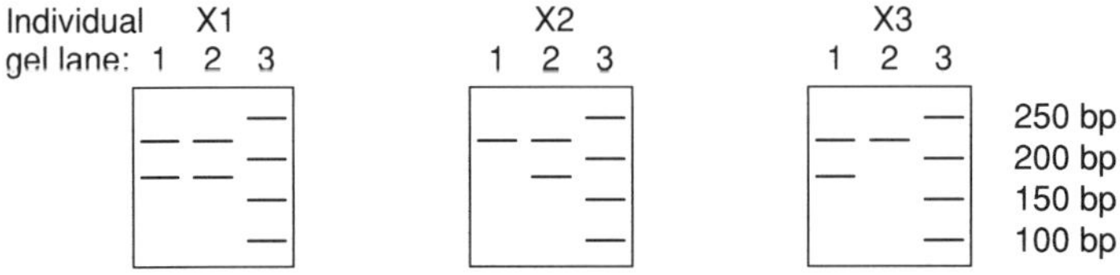

Primer 1 (common upstream): 5′ TGT TCC GTC CCT GGA GAT TAT 3′
Primer 2: (normal allele specific): 5′ T TGA CTT CAT ACT GAA GGC A**C** 3′
Primer 3: (mutant allele specific): 5′ T TGA CTT CAT ACT GAA GGC A**G** 3′
Primer 4: (internal control): 5′ CAT CGT TTC CTG CTC TTT AAA 3′

Multiplex PCR reactions
Reaction 1: genomic DNA sample, and primers 1, 2 and 4
Reaction 2: genomic DNA sample, and primers 1, 3, and 4

Gel electrophoresis
Lane 1: load a sample of PCR reaction 1 for the individual
Lane 2: load a sample of PCR reaction 2 for the individual
Lane 3: load a sample of the molecular weight standard

Partial sequence of the normal allele (B) *(numbers are nucleotides; 1269 is a T)*

```
                                          1269                                  1296
3' TTG ACA AGG CAG GGA CCT CTA ATA GGT AAA ATG AAG TTA CTT TAA GTC AGT TGT
5' AAC TGT TCC GTC CCT GGA GAT TAT CCA TTT TAC TTC AAT GAA ATT CAG TCA ACA
   N   C   S   V   P   G   D   Y   P   F   Y   F   N   E   I   Q   S   T
                    Primer 1

                                          1323                                  1350
   TCA CTG TAG TAA CTT CCT TTA ATA CCA CCA GTT CAC CTC TTT GAG TAG ATG CCA
   AGT GAC ATC ATT GAA GGA AAT TAT GGT GGT CAA GTG GAG AAA CTC ATC TAC GGT
   S   D   I   I   E   G   N   Y   G   G   Q   V   E   K   L   I   Y   G

                                          1377                     Primer 2     1404
   CAG AAG TGC TGT GGT CAC TTG AGA TAA CCA CCG AGA CGA CCA ACA CGG AAG TCA
   GTC TTC ACG ACA CCA GTG AAC TCT ATT GGT GGC TCT GCT GTT TGT GCC TTC AGT
   V   F   T   T   P   V   N   S   I   G   G   S   A   V   C   A   F   S

                                          1431                 Primer 4         1458
   TAC TTC AGT TAT GAA CTC AGT AAA CTA CCA GGT AAA TTT CTC GTC CTT TGC TAC 5'
   ATG AAG TCA ATA CTT GAG TCA TTT GAT GGT CCA TTT AAA CAC CAG GAA ACG ATG 3'
   M   K   S   I   L   E   S   F   D   G   P   F   K   E   Q   E   T   M
```

6. The following DNA chip sequencing of normal and mutant coding strand at a particular locus reveals what mutation in the mutant's coding strand?

normal

	1	2	3	4	5	6	7
A	●	○	○	○	○	○	○
C	○	●	●	○	○	●	○
T	○	○	○	○	●	○	●
G	○	○	○	●	○	○	○

	1	2	3	4	5	6	7
A	○	○	○	○	○	○	○
C	○	○	○	○	○	○	○
T	○	○	○	○	○	○	○
G	○	○	●	○	○	○	○

A A deletion of six bases
B The insertion of one base
C The substitution of a C for a G
D A deletion of six bases and the addition of one base
E None of the above

Review questions

1. What is the difference between genetic testing and gene tracking? What methods are used more for one than the other?
2. How can one screen for a specific mutation?
3. How can one screen for mutations in a specific gene?
4. What is SSCP analysis?
5. What is the PTT?
6. How are ASOs used in genetic testing?
7. What is the ARMS test?
8. What is RFLP analysis?
9. How can DNA arrays be used in genetic testing?
10. How can DNA arrays be used in sequencing genes?

Open-ended problem

1. Describe how the ARMS test could be used to detect a disease allele at the Tay Sachs locus.

18 Cancer genetics

Multiple choice questions

1. Which of the following best describes the two-hit hypothesis?

A An allele of a gene must be mutated twice to initiate tumorigenesis
B Two gene loci must be mutated to initiate tumorigenesis
C Two cellular proteins must be defective to initiate tumorigenesis
D Both alleles of a gene must be mutated to initiate tumorigenesis
E None of the above

2. Which of the following is NOT a known class of oncogenes?

A Transcription factors (e.g. AP-1)
B RNA helicases (e.g. RH-2)
C Tyrosine kinases (e.g. ABL)
D Growth factor receptors (e.g. EGF-R)
E All of the above are examples of known oncogenes

3. Which of the following techniques is the best suited for identifying potential oncogenes?

A Developmental Northern blot
B NIH-3T3 assay
C Affinity chromatography with resin conjugated with tumor cell-derived DNA fragments
D Inverse PCR using tumor cell-derived DNA and primers designed to exons in the tumor genome
E None of the above are well suited for the task

4. Tumor DNA and normal control DNA were labeled with green and red fluorescent labels, respectively, then hybridized *in situ* together in equal quantities to chromosomes of normal cells. The curve below shows a computer-generated scan of the ratio of green:red fluorescence intensity along the chromosome. How do you interpret this particular data?

A The tumor has an amplification of DNA at 8q24.1
B The tumor has a deletion at 8q24.1
C The tumor has increased expression of genes at 8q24.1
D The tumor has decreased expression of genes at 8q24.1
E None of the above

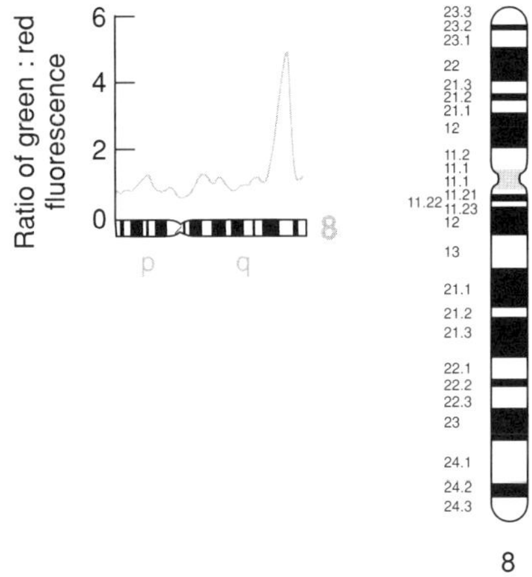

5. Which of the following genes when mutated directly increases the number of DNA changes by lowering the fidelity of the DNA replication process?

A Mutator genes
B Tumor supressor genes
C Oncogenes
D Two of the above
E All of the above

Review questions

1. Describe the difference between a tumor suppressor gene, a mutator gene, and an oncogene. Provide an example of each, indicating gene function in addition to gene name.

2. Distinguish between the two-hit hypothesis of tumor formation and the multi-step model of carcinogenesis.

3. What is the normal function of p53? How does the loss of p53 lead to an increased susceptibility to cancer?

4. What is the difference between a normal cell, a tumor cell, and a cancer cell?

5. How can viruses contribute to cancer?

6. What are examples of oncogenic mutations?

Open-ended problem

1. If a tumor either has zero, one or two copies of a given tumor suppressor gene, why is it that one often sees only a *partial* loss of one allele's band on a gel when analyzing pathological tumor samples?

19 Studying human disease genes

Multiple choice questions

1. You've recently joined a group interested in tissue-specific adhesion molecules. You suggest to your advisor that it would be worthwhile to characterize the regulatory region of one of the genes she recently identified. She wants to know the parts of the region essential for regulating the gene's expression in cardiac muscle. You confidently tell her you can find that out *down to the base pair* for all the factors involved. How would you do this?

A Do a primer extension experiment to identify the start of transcription and then identify the TATA box. Show her where the RNA pol II binds based on consensus binding sites
B Do nuclease S1 protection assays allow you to identify the start of transcription and, hence, the regulatory domain. You can find SP2 binding sites by visual inspection or computer-based pattern searches
C Do a DNA footprinting assay with cardiac cell nuclear extract. See what part of the regulatory sequence is protected from digestion in cardiac cells compared to others
D Do affinity chromatography with the regulatory domain and muscle cell lysate; purify DNA binding proteins. Sequence the N terminus of the proteins; clone them
E None of the above

2. The figure below shows the results of a luciferase assay performed on cells that have been transfected with plasmids in which fragments of the regulatory region of gene X have been cloned in front of the luciferase gene. Which of regions between –200 bp and the start of transcription (+1 bp) appears to contain binding sites for transcription inhibitors?

A –200 to –150
B –150 to –100
C –100 to –50
D –50 to 0
E The data don't reveal any inhibitory sites

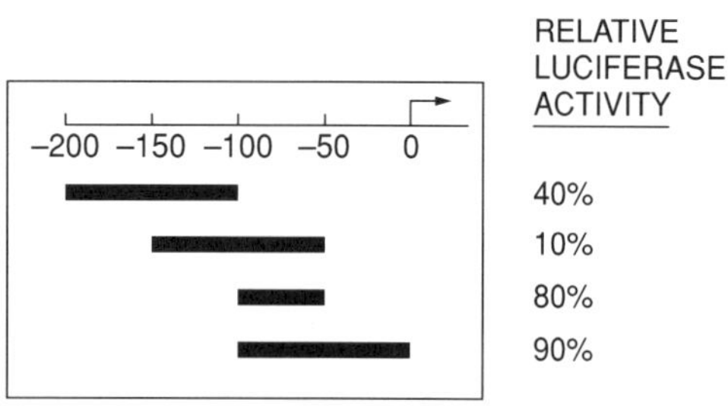

3. The Northern blot shown below has had an equal amount of RNA loaded in each lane. RNA came from the different stages and adult tissues indicated at the top of each lane. Which of the following does one observe in this data?

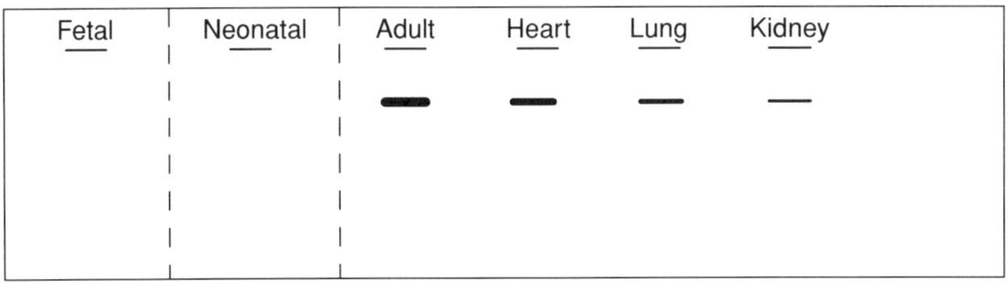

A Embryonic expression
B Adult expression
C Stage-specific alternative RNA splicing
D Heart-specific expression
E None of the above

4. An inverse PCR experiment intended to reveal exon–intron boundaries, using cDNA specific primers yielded the following two sequences (not all sequence obtained is shown) when the same primers that were used for the inverse PCR were used for sequencing. The sequence shown is NOT the sequencing *template* DNA but the sequence read directly off the dideoxy sequencing gel. Which of the exon–intron boundary results (boundaries marked with '|') are consistent with this sequence data?

Sequence 1: 5′ (reverse primer sequence) ... TGCC | CTGA ... 3′
Sequence 2: 5′ (forward primer sequence) ... TCAG | GTAA ... 3′

A 5′ TCAG | GTAA ... intron ... TCAG | GGCA
B 5′ TCAG | GGCA ... exon ... TCAG | GTAA
C 5′ TCAG | GGCA ... exon ... AATG | GACT
D 5′ ACGG | GACT ... intron ... AATG | GACT
E None of the above

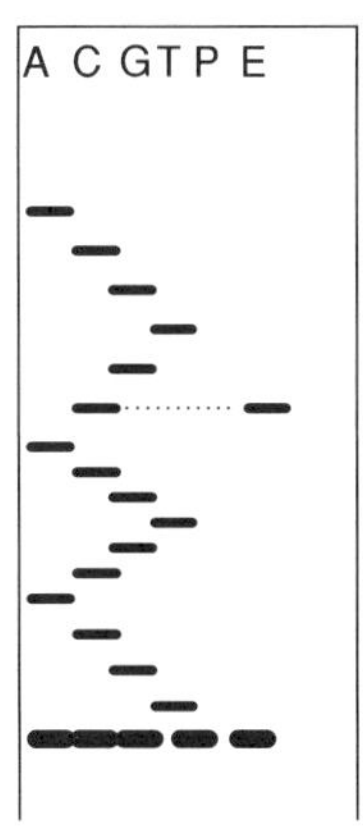

5. A primer extension experiment with a labeled primer downstream of the start of transcription, oriented to allow synthesis of the template strand (i.e. synthesis toward the start of transcription) was performed and the radiolabeled product run on the gel alongside a sequencing reaction using a genomic DNA fragment and the same primer. Which base is the first base added during RNA transcription?

A A
B C
C G
D U
E Cannot be determined from this data

6. A 5′ RACE–PCR procedure was performed using a primer 1900 bp upstream from the stop codon of a gene and mRNA from a tissue expressing the gene. The product generated was 50 bp long. Assume a 100 bp polyA tail is added to the gene's transcript 50 bp downstream of the stop codon. How long is the open reading frame of this gene?

A 1950 bp
B 2000 bp
C 2050 bp
D 2100 bp
E Cannot be determined from the information given

Review questions

1. Explain how two people of medium height could have a child who grows to be much taller than either parent *without* having had any new mutation or any difference in diet or developmental environment? If this happens does it argue against the concept of regression to the mean?

2. What is Falconer's polygenic model for dichotomous nonmendelian characters? How might it explain a disease such as schizophrenia?

Open-ended problems

1. In a comprehensive survey of physicians throughout a region with excellent primary health care, 40 families are located with children affected by a serious but non-lethal disease. Of these families, 33 were families in which there were only two children. Of these familes, 28 had one child affected and 5 had two children affected.

   (a) What is the true, unbiased segregation ratio of this disease?

   (b) Of families with *three* children in which one or more is affected, what would you expect would be the relative frequency of the following: families with one child affected, with two children affected, and with three children affected?

20 Studying human gene structure

Multiple choice questions

1. The main difference between nuclease S1 protection and primer extension analysis is

A the need for DNA sequencing following the nuclease S1 protection assay
B the use of DNA molecules in the primer extension analysis
C the requirement of an RNA dependent DNA polymerase in primer extension analysis
D labeling (often isotopic) of the DNA fragment in the nuclease S1 protection assay
E none of the above are differences

2. To perform a BLAST search using a 50 amino acid protein sequence against the DNA contents of GenBank translated in six frames, which search program would one use?

A BLASTN
B BLASTX
C BLASTP
D TBLASTN
E EST BLAST

3. What is the advantage of Northern hybridization over *in situ* hybridization, quantitative RT–PCR, and microarray hybridization?

A Northern hybridization is more sensitive than all these
B Only Northern hybridization will provide transcript size
C Only Northern hybridization can tell one which cells make the transcript
D Northern hybridization has the highest throughput
E None of the above

4. Serial analysis of gene expression

A is a method for multiplex gene expression
B depends on very short sequence tags located at specific sites within individual transcripts
C is very fast; *all* genes expressed at 0.5% of total mRNA in one tissue but not abundant in another could be found in a *single* day
D provides a measure of transcript quantitation since genes that are expressed strongly in a specific tissue will frequently be recovered as sequence tags
E All of the above are true

5. The yeast two-hybrid system includes all of the following except

A creation of two strains of hybrid yeast, one strain that contains a gene for one fusion protein, the other strain that contains a second fusion protein
B the effective reconstitution of a transcription factor, which normally contains a DNA binding domain and an activation domain, using the interaction of peptides (or RNA and peptide) that are fused to the transcription factor domains
C a reporter gene that reveals the binding of an interacting protein (fusion) to the bait and that fusion complex's promoting the reporter's transcription
D can be used to investigate protein–protein networks in yeast
E fusion vectors allowing hybrid proteins to be made

Review questions

1. How do you obtain full-length clones of human genes?
2. How do you determine the start of transcription of a gene?
3. How do you determine exon–intron boundaries?
4. How can you determine the actual N-terminus of a protein for which you have sequenced the cDNA?
5. How can you determine when, where, and how highly a gene is expressed?
6. How can you determine whether a regulatory sequence is involved in promoting or inhibiting transcription?
7. How can you determine which proteins can bind to a piece of DNA?

Open-ended problems

For each question below, briefly describe two technical approaches one might use to answer. Your response should include the name of the procedure and a description of the procedure's essential steps.

1. You have obtained a cDNA for a gene encoding a gene upregulated in metastatic breast cancer cells. What is the complete sequence of the mature transcribed mRNA?

2. You know the approximate transcription start site for the gene you are studying. What is the exact nucleotide that is first added by RNA polymerase when this gene is expressed?

3. Once you have the complete cDNA sequence in hand, how do you determine the exons that were spliced together during mRNA processing? What use is there in knowing this?

4. If the composite (full-length) cDNA sequence you have obtained has three ATGs at the beginning of the longest ORF, how do you know which one(s) is used as translational start sites in the organism?

5. Does your cDNA encode a glycoprotein? Or a phosphoprotein? Or a secreted protein? Or a nuclear protein? How might you use biological databases to obtain immediate insight into the function of the gene you have cloned?

6. When during an individual's lifetime is the gene you have cloned expressed? In what tissues is it normally expressed? Is the gene's expression reduced upon treatment with any of the existing drug therapies? Do expression levels vary from cell type to cell type? Is the gene alternatively spliced in different cell types or in tumor cells?

7. What enhancer or inhibitory elements are upstream of the gene coding sequence?

8. In normal breast tissue, what upstream regions of the gene you have identified do proteins bind to? How could you identify the proteins that bind?

21 Genetic manipulation of animals

Multiple choice questions

1. Embryonic stem (ES) cell transplantation

A avoids the problem of chimerism of germline cells associated with oocyte injection
B was used in the cloning of Dolly
C allows one to select for the rarely occurring homologous recombinant cells before injection into a blastocyst cavity and implantation into a surrogate mother
D requires the isolation of inner cell mass from a meiotically arrested oocyte for transplantation
E All of the above

2. YAC transgenics

A can be generated by spheroplast fusion with embryonic stem cells
B have permitted study of large genes which cannot be transmitted by plasmid vectors
C were used to create mice that can produce human antibodies
D can be used to model diseases caused by large scale gene dosage imbalance
E All of the above

3. All of the following are true of the Cre-LoxP system EXCEPT it

A can mediate either intrachromosomal or interchromosomal recombination
B artificially generates site-directed recombination
C uses a recombinase derived originally from a bacteriophage
D can allow one to make conditional knock-outs
E allows one to create recombinants with low x-over (cross-over) efficiency

4. The knock-in method allows one to

A select against nonhomologous insertions using gancyclovir treatment
B knock-out a gene with another coding sequence that is then regulated by the upstream sequence of the knocked-out gene
C study tissue-specific expression patterns of the gene of interest
D have a selectable marker (such as *neo*) that can be removed by crossing
E All of the above

5. **Gene trapping**

A is a technique that involves chromosome walking or jumping followed by rapid transcript analysis
B identifies expressed genes in genomic sequence by computer recognition of consensus sequences for promoters, splice sites, and polyadenylation sites
C identifies exons in genomic sequence by using a vector containing exons flanking the multiple cloning site
D uses an expression defective transgene to select for chromosomal integrations that occur within or close to genes
E None of the above

6. **Dolly the cloned sheep has which of the following?**

A Nuclear DNA derived entirely from her clonal mother
B A chimeric germline
C A quarter of her clonal mother's father's chromosomes
D A serum-depleted denucleated oocyte mammary in her immediate ancestry
E None of the above

Review questions

1. How do you make a transgenic mouse by oocyte injection?
2. How do you make a transgenic mouse by ES cell transplantation?
3. How are YAC transgenics made?
4. What is the Cre-LoxP system of mutagenesis?
5. How is double replacement gene targeting done?
6. How is a knock-in mutation made?
7. How is a knock-out mutation made?
8. What is gene trapping?
9. How does one clone an animal?

Open-ended problems

1. Outline the steps required to generate a LIMK1 knockout in mice by homologous recombination.

2. How could you use the Cre-LoxP system to inactivate the dopamine receptor 4 (DR4) specifically in the substantia nigra?

22 Gene therapy and other molecular genetic-based therapies

Multiple choice questions

1. **Of the following gene therapy vectors, which has a high efficiency of DNA transfer, a somewhat low efficiency of *integration* into the genome, a tropism for the respiratory epithelium, and a tendency to cause an immune response?**

A Adenovirus
B Retrovirus
C Herpes virus
D Liposomes
E None of the above

2. **Several characteristics contribute to a disease condition being particularly well suited for gene therapy approaches. Which of the following diseases is *best* suited for a gene therapy application?**

A Huntington's disease
B Down syndrome
C Adenosine deaminase deficiency
D Schizophrenia
E None of the above

3. **There are many potential challenges and hazards to current gene therapies. Which of the following is NOT one of them?**

A Immune reaction to gene therapy vector
B Delivery of gene therapy product to target tissue
C Loss of gene therapy product expression over time
D Radiation damage to tissue during gene therapy delivery
E Disruption of critical gene by integration of gene therapy vector

4. Modifying a patient's own cells in a petri dish (or plastic flask) by adding a gene therapy vector carrying a human gene that is homozygous for the mutant form in the patient, and then *placing those cells that took up the vector back into the patient* would be called an __________ gene therapy

A *in vivo*
B *in vitro*
C *ex vivo*
D *ex vitro*
E None of the above

5. Which of the following is NOT an established approach to gene therapy?

A Targeted mutation correction
B Assisted killing of disease cells by immune system cells
C Gene augmentation therapy
D Replacement of mutated genes with normal genes in all somatic cells
E All of the above are standard approaches to gene therapy

Review questions

1. What are the features of a disease or condition that makes it especially suitable for gene therapy approaches?

2. Discuss the advantages and disadvantages of three different gene therapy delivery vectors. How can the disadvantages of each be reduced?

3. Describe the differences among dominant negative protein inhibitor, antisense, and catalytic RNA(ribozyme)-based gene therapy strategies.

4. Explain three broadly different ways to direct the expression of gene product delivered to cells in a gene therapy protocol.

Open-ended problem

1. Gene therapy is an emerging field of medicine, with many failures, several sizable steps forward, and few dramatic successes. If you were interested in developing a gene therapy for a particular condition, what questions related to the design, execution, and analysis of the work would it be important to address before beginning? There are many!

1 DNA structure and gene expression

Answers to MCQs

1. C 2. B 3. E 4. C 5. D 6. A 7. D 8. B 9. D 10. C

Review

1. See diagram (Watson–Crick base pairs, annotated) below.

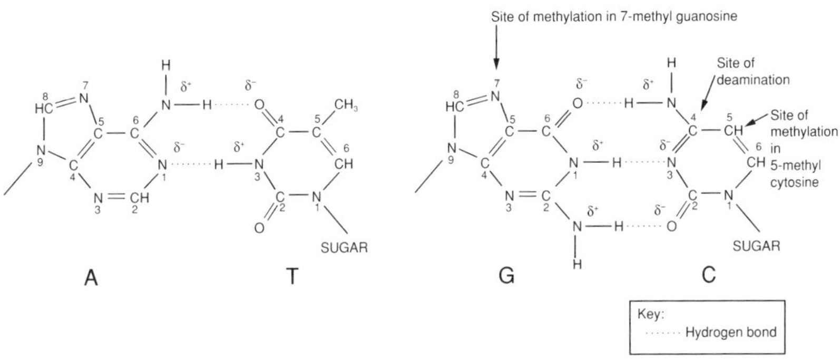

2. (a) Acidic: Asp, Glu.
Basic: Lys, Arg, His.
Uncharged polar: Asn, Gln, Ser, Thr, Tyr, Cys.
Nonpolar neutral: Gly, Ala, Val, Leu, Ile, Pro, Met, Phe, Trp.
(b) Largest side groups: Trp, Tyr (followed by Phe and Arg).
(c) Forms disulphide bridges: Cys.
(d) Encoded by 5′ ATG 3′: Met.

3. (a) The TATA box is first bound by TFIID and later by TFIIA. GC boxes (5′ GGGCGG 3′) are bound by ubiquitous transcription factor SP1.
(b) The promoter is the sequence just upstream of the coding sequence (for genes transcribed by RNA polymerase II) that is bound by a complex of RNA polymerase and general transcription factors at the initiation of transcription. It is the *cis*-acting element closest to the coding sequence.
Enhancer sequences are *cis*-acting elements near the promoter, or far upstream or downstream from the coding sequence, that are *not* bound by the

RNA polymerase but by transcription factors that can act at a distance to enhance the rate of transcriptions.

(c) Antiparallel refers to the fact that for most nucleic acids, strands pair with strands with the opposite polarity (running in the opposite direction): 5′ to 3′ binds to 3′ to 5′.

Complementarity refers to the rules of base pairing (A:T and C:G). Two antiparallel sequences needn't be completely complementary, but two complementary sequences will be so only if they are in an antiparallel alignment.

(d) The splice junction is either between the end of an exon and the beginning of an intron or between the end of the intron and the beginning of an exon. These mark the bases of the exons joined together during splicing. The branch site is within the intron, and marks the site at which the 5′ end of the intro binds to the internal base of the intron to form the lariat structure.

Open-ended problems

1. See diagram below (annotated splice intermediate).

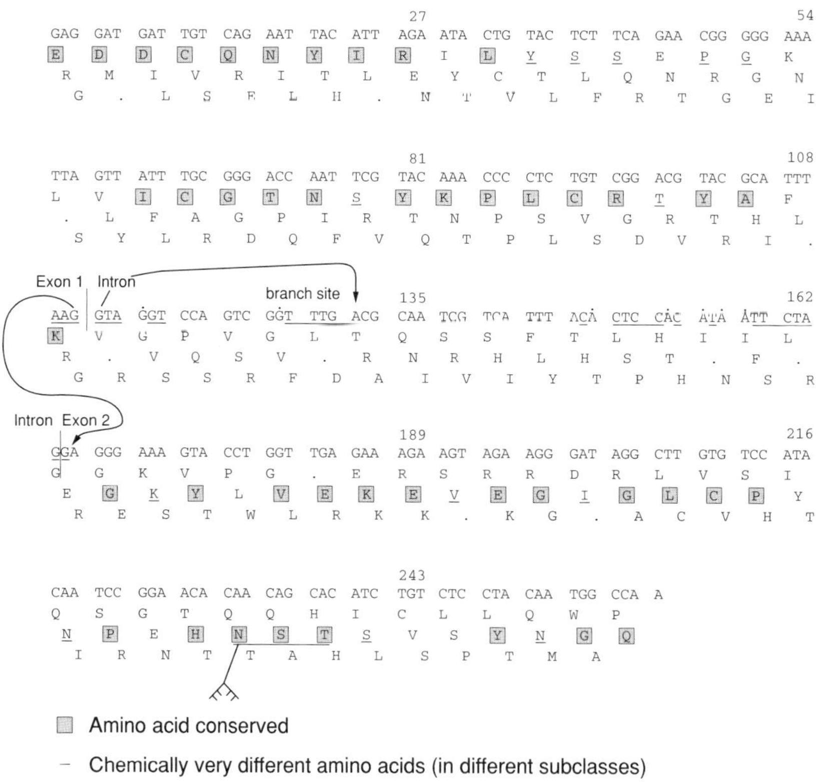

2. Note that by convention the 5′ to 3′ strand is written above the template strand if the template strand is shown at all.

(a) 5′ ACUUUCCGUA ... 3′

(b) 7-Methyl guanosine would be attached via a 5′ to 5′ linkage to the first A in the transcript. The A and what follows it would also have a methyl group instead of a hydrogen atom on the oxygen that is attached to the 2′ carbon of their ribose moieties.

(c) There are many GC boxes (GGGCGG) in the regulatory region of *SP1*, including those portions that are 'protected' by proteins in the nuclear extract. SP1 binds to GC boxes. Hence, protein SP1 appears to be partly autoregulated.

(d) *SP1* is an example of a gene that lacks a TATA box. The initiation of transcription of the *SP1* gene doesn't require TATA-binding proteins. As a housekeeping gene, *SP1* is always needed and so fits the profile of many of the TATA-less genes.

2 Chromosomes in cells

Answers to MCQs

1. D 2. A 3. D 4. B 5. D 6. E 7. D 8. C 9. D 10. D

Review questions

1. Hepatocytes. Platelets are nulliploid. Typically methylation increases with differentiation.

2. n and C for various organisms has been determined. See HMG2, Table 2.1.

n for mouse = 20 (chromosomes in the haploid set)
C for mouse = 3.5×10^{-12} g (there are 3 000 000 000 bp in the haploid set)
n for onion = 8 (chromosomes in the haploid set)
C for onion = 1.05×10^{-11} g (with 15 000 000 000 bp, onion has 3× the C value)

3. (a) Coiling or folding the DNA in a regular manner (condensation to the 30 nm fiber mediated by histone tails and histone phosphorylation).
(b) Protecting the DNA from nuclease activity (especially DNAse I; sensitivity correlates with localized decondensation).
(c) Chromatin remodeling (generally as a remodeling repressor; repression released by chromatin-remodeling complexes).
(d) Regulation of gene transcription (generally as gene transcription repressor; repression released by histone acetylation).
(e) Regulation of heterochromatin formation.

4. The three essential components of a eukaryotic chromosome are the following (with examples of molecular make-up of each given for *yeast:*

(a) Origins of replication. There must be many of these per chromosome to ensure that DNA is replicated in the time available.

```
ATTTAT(A/G)TTTA
TAAATA(T/C)AAAT
```

and several other imperfect copies of this.

(b) A centromere to ensure the proper segregation of chromosomes during mitosis and meiosis.

```
TCACATGAT<---80-90 bp of >90% (A+T)--->TGATTTCCGAA
AGTGTACTA                              ACTAAAGGCTT
```

(c) Two telomeres to protect the ends, and ensure that they are replicated completely.
$(TG)_{1-3}$ T G_{2-3}
$(AC)_{1-3}$ A C_{2-3}

5. Mixoploidy is the presence of more than one chromosome number in a cell population. It can accompany two genetic states: mosaicism and chimerism. In the case of mosaicism, a chromosome is lost or gained during development of an organism such that some proportion of the cells in the organism are euploid, and some are aneuploid. All the individual's cells, though, are derived from the same fertilized egg.

In the case of chimerism, an individual is composed to two distinct cell lines each derived from a separate fertilized egg. Two zygotes or embryos fuse giving rise to one chimeric individual. If one of those two zygotes or embryos is aneuploid, the resulting chimera will be mixoploid.

6. In G-banding, the Giemsa dye preferentially stains the AT rich (gene-poor) heterochromatic regions of the partially trypsinized chromosomes.

With R-banding, the chromosomes are first heat-denatured in saline, so that AT rich regions are more open (and less capable of holding the Giemsa dye). GC regions sequester more of the dye. Sometimes a dye that is GC-specific is used to generate G bands.

7. The function of each of these procedures is as follows.
(a) Addition of phytohemagglutinin (a lectin, or carbohydrate binding protein) induces T lymphocytes, a class of white blood cells, to divide. Cells going through mitosis will have condensed chromosomes. Cells in interphase won't.
(b) Thymidine starvation will prevent cultured cells from proceeding through the cell cycle. When thymidine is again put in the culture, cells will begin cycling *synchronously*, allowing one to capture a high proportion of the cells in prophase if one times the preparation correctly.
(c) Colcemid prevents the formation of microtubules that would normally form the spindle fibers in dividing cells. Without the ability to form a spindle, the cells can't proceed from prophase to metaphase. The result is more cells in the preparation will be in prophase. The mitotic index will be higher.
(d) Treatment of cells with hypotonic saline causes them to swell. This improves the informativeness of the spread of chromosomes generated from the cells.
(e) Trypsinization of chromosomes partially degrades the protein scaffolding that condenses chromatin, preventing access to dye. Careful regulation of the extent of trypsinization allows differential opening of AT rich regions, which allows bands to be seen.
(f) Heat-denaturization of chromosomes in saline makes the AT-rich regions less attractive to the dye, making possible the preferential staining of R bands.
(g) Giemsa staining involves partial trypsinization of chromosomes followed by addition of Giemsa dye which under these conditions binds preferentially to the AT-rich regions of chromosomes.
(h) Quinicrine, DAPI and Hoechst 33258 all preferentially bind to AT-rich DNA. These dyes are fluorochromes, requiring an ultraviolet wavelength excitation energy to produce fluorescence in the visible range.

(i) Chromomycin A3, olivomycin, and mitramycin are GC-specific dyes often used in G-banding procedures.
(j) Saturated solutions of barium hydroxide denature DNA specifically at sites such as centromeres that are thought to be constitutionally heterochromatic.

Open ended problems

1. The ways of identifying the centromere in a mammalian chromosome are
   (a) Look for the primary constriction in metaphase chromosomes using a microscope.
   (b) Perform C-banding of the chromosomes by treating cells with barium hydroxide.
   (c) Stain with antibodies to centromere binding proteins (such as those in the kinetochore) followed by anti-idiotype fluorescently labeled secondary antibodies.
   (d) Use biotinylated DNA probe for hybridization to centromere sequences.
   (e) Look for late DNA synthesis in chromosomes in meiosis II.

2.

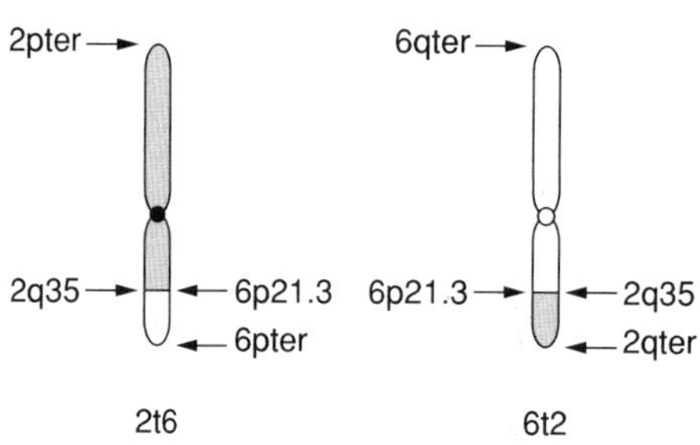

3. 46, X, X t(7,14) (q31.3;q21)

   The abnormality is a reciprocal translocation.

4. Y ml blood × 5000 lymphocytes/ml blood × 2 haploid genomes/lymphocyte (diploid) × (3.5×10^{-12} grams of genomic DNA/haploid genome × (1×10^{6} micrograms/gram) = 100 micrograms of genomic DNA from lymphocytes

   Y = 2857 ml of blood (almost three liters!)

5. In a high-resolution banding of a human chromosome spread there are 850 bands. 100 000 genes, spread evenly among these bands would suggest about 118 genes per band. Assuming, however, that most of the genes will be found in R bands, not G bands, then an average sized R band would have up to about 235 genes. Obviously, with a less skewed distribution of genes between R and G bands this latter number would drop. Larger than average R bands would be expected to have a proportionally greater gene content. One would also expect that there may be some particularly gene-rich parts of the genome and some

particularly gene-poor parts of the genome that would spread the expected range further.

With about 3 billion base pairs and 850 bands per haploid genome, one would expect an average of about 3.53 million bases per average sized high-resolution band.

3 Genes in pedigrees

Answers to MCQs

1. C 2. C 3. A 4. D 5. E 6. E 7. D 8. A 9. E 10. A

Open-ended problems

1. (a) http://www.ncbi.nlm.nih.gov/Omim/ is the universal resource locator (URL) of OMIM. Simply type that (exactly) into the 'go' to box at the opening screen of your browser and hit return. Search for a disease or other character by entering the name into the search box and then choose one of the hits that the search uncovers to go to that listing.

(b) Read the Clinical Synopsis or the text section of the listing to glean the essence of the mutant or disease phenotype, in contrast to wild-type.

(c) Usually dominance or recessiveness will appear fairly prominently in the text.

(d) Complexities might include, for example, partial penetrance, variable expressivity or anticipation.

2. Several hypotheses could be proposed, including the following

(a) The two people may have noncomplementing deafness alleles (i.e. their deafness is caused by recessive alleles at the same locus).

(b) The one child who is not deaf could have had a spontaneous reversion mutation.

(c) The deafness trait may be only partially penetrant.

(d) The child who is not deaf has not yet reached the age of onset – perhaps variable.

(e) The parents may be Aabb and aaBb and so give rise to an occasional AB child.

3. Given the low likelihood of paternal mitochondrial transmission, new mutation is the more likely of the two. Still, comparing the father's mitochondrial DNA and the child's for shared RFLPs (by restriction enzyme digestion and Southern analysis) or sequence polymorphisms (by dideoxysequencing of the polymorphic D-loop) would quickly resolve the question.

4 Cell-based cloning

Answers to MCQs

1. C 2. C 3. A 4. B 5. B 6. E 7. D 8. C 9. B 10. E

Review questions

1. Type II restriction endonucleases cut within their recognition sequence and therefore leave sticky ends (if they are not blunt cutters) that can be ligated to similar ends yielding a regenerated restriction site generally able to be recut by the same restriction enzyme.

2. The advantages and disadvantages of the vectors are

   (a) pUC-based plasmids
   Advantages: (1) ease of manipulation, transformation, and preparation; (2) can carry small inserts such as cDNAs; (3) can be modified in many ways to increase usefulness (e.g. with bacterial promoters adjacent to multiple cloning site).
   Disadvantages: (1) small insert size (seldom greater than 5–10 kb inserts); (2) relatively inefficient efficiency of transformation.

   (b) Lambda phage
   Advantages: (1) can carry larger inserts (0–23 kb depending on the lambda phage vector used); (2) well-suited for carrying large cDNAs or small segments of genomic DNA; (3) efficient infection of bacterial hosts; (4) easy to screen bacteriophage plaques.
   Disadvantages: (1) limited insert capacity; (2) more complicated propagation techniques.

   (c) Cosmids
   Advantages: (1) can carry inserts of 30–44 kb; (2) can be maintained episomally.
   Disadvantages: (1) tendency to recombine can lead to rearranged inserts and fusions between sequences not normally adjacent to each other; (2) once a cosmid is prepared from host, it is more difficult to modify it and return it to host again than a plasmid is (partly because of its size).

   (d) Bacteriophage P1
   Advantages: insert capacity of around 70–100 kb more than doubles that of cosmids.
   Disadvantages: more complicated to work with.

(e) YACs

Advantages: huge insert capacity (200 kb to 2 MB).

Disadvantages: (1) large size makes construction, transformation, and purification very cumbersome; (2) structural instability (due largely to the presence of repetitive sequence) can lead to rearrangement of cloned DNA.

Open-ended problems

1. **Steps**
   - Prepare RNA from lymphocytes that have been treated with the cytokine; this will induce the transcription of the gene encoding the cytokineX-inducible factor (CXIF) gene.
   - Prepare cDNA from this RNA with the use of reverse transcriptase for the first strand synthesis and DNA polymerase for the second strand synthesis.
   - Attach linkers (containing restriction sites) to the ends of the cDNAs.
   - Ligate cDNAs into expression plasmids (behind an inducible promoter).
   - Transform bacteria with these plasmids.
   - Plate bacteria on agarose plates with an antibiotic that prevents untransformed bacteria from growing and with the inducer molecule that results in the expression of the cDNA cloned into the expression vector.
   - Place disks of nitrocellulose onto plates so that samples of bacteria (and the proteins made therein) will be transfered to them (a solid porous surface).
   - Fix proteins and DNA to the disks by UV crosslinking.
   - Incubate disks with a non-specific blocker and with an appropriate dilution of the antibody that will recognize the CXIF.
   - Wash and incubate disks with a secondary antibody that will bind to the anti-CXIF antibody and bring to it a chromophore (colored compound or enzyme that generates such).
   - Identify the colony(s) corresponding to the colored spot(s) on the nitrocellulose filter.
   - Repeat the plating and 'probing' steps described above until the colony is clonal (not contaminated by any bacteria not containing the same plasmid).
   - Prepare plasmid DNA from the plasmid; sequence the insert.
   - Verify that insert encodes a protein that is
     (a) secreted/membrane bound (signal sequence)
     (b) recognized by the anti-CXIF antibody
     (c) and is transcribed (and/or translated and/or delivered to the surface) at higher levels in cells treated with the cytokine than when not.

   Assumptions
   - The bacterial protein will be recognized by antibody (i.e. antibody doesn't recognize a carbohydrate epitope on the protein that is not added by the bacteria).
   - The protein is not so big that a cDNA bearing the epitope will never be found in a cloned plasmid.
   - The fixation procedure does not denature the protein in such a way that the antibody will not recognize it.

   You might also have suggested purifying the protein from lymphocytes using affinity columns using the antibody and then microsequencing the protein, designing degenerate oligonucleotide probes or primers and screening cDNA libraries with these. That would be a very difficult path if the protein were of very low abundance as might well be the case.

2. Chromosome walking would be one way to go after the *mbm-B* gene with your probe.
   - Isolate genomic DNA from isogenic wild-type flies.
   - Make a genomic library by inserting partially digested fragments of this DNA into lambda phage vector.
   - Package the library and titer it [determine how many plaque forming units (p.f.u.) there are per microliter].
   - Infect bacteria with the phage and plate phage on agarose plates.
   - Place disks of nitrocellulose on the plates once large plaques have formed.
   - Crosslink DNA/protein onto filters (with UV light or by baking).
   - Probe filters with 200 bp piece of DNA labeled radioactively or otherwise.
   - Identify hybridizing genomic clones and prepare DNA from them.
   - Probe cDNA libraries (made as in problem 1 above) with the various genomic fragments.
   - Check to see if any of the cDNAs hybridize to embryonic (or larval or adult) brain tissue, specifically in or near the mushroom bodies/olfactory lobes.
   - If so check *mbmB* homozygotes for alteration in abundance, size, or sequence of that cDNA.
   - If not, make a restriction map of the genomic fragments obtained and probe the genomic library with probes made from fragments at the termini of the clones.
   - Repeat above steps until a cDNA is found that is
     (a) expressed in the expected tissue (assumptions!)
     (b) maps cytologically to the expected band
     (c) and has an altered expression or sequence in the *mbm-B* mutant.

3. (a) *Bsm*I should be identified at base 791. *Fok*I should be identified at both 533 and 678. In both cases, the bond broken should match the location of the caret (^) in the restriction site list.
   (b) An isoschizomer.
   (c) *Sna*BI at 7bp and *Xba*I at 1000 bp (releasing 6 and 20 bp pieces).
   (d) Type I.
   (e) Yes.
   (f) No. *Sal*I has a 3′ AGCT 5′ underhang. *Taq*I has a 3′ GC 5′ underhang.
   (g) (940 – 79) bp insert + (4887 bp – 51 bp) = 5697 bp.
   (h) *Spe*I and *Xba*I produce compatible ends. Therefore the vector and insert will preferentially ligate in <u>intra</u>molecular reactions instead of the desired intermolecular one. Vector dephosphorylation would help reduce the vector-only ligation but would not prevent the insert-only intramolecular ligation. Also, the cloning would occur with the insert in both orientations, only one of which might be desired.
   (i) This answer depends on what you assume the fraction of A, C, G, and T are in the genome. Here you can assume 25% for each. At a given location the probability of G, $p(\mathrm{G})$, is 0.25, $p(\mathrm{C}) = 0.25$, $p(\mathrm{A}) = 0.25$, $p(\mathrm{T}) = 0.25$. So, $p(\mathrm{GTAC}) = (0.25)(0.25)(0.25)(0.25) = (0.25)^4$, which gives us 4 sites predicted to be found in 1080 bases (4.22 to be more precise).

(j) $p(\text{GCGC}) = (0.25)(0.25)(0.25)(0.25) = (0.25)^4$ or 4 in 1080 bases. There are 5 GTAC sites and no GCGC site. One reason for the discrepancy is that GCGC sequences have functional significance (as CpG islands have in mammals) and may therefore have been selected against in this coding sequence.

4.
- Mobilize P-elements in flies so that they insert in germline tissues into new sites in the genome. Introduce a source of *transposase* into flies by injection of embryos or by crossing with a strain of fly that carries a functional copy of the transposase gene.
- Raise progeny and select for those that have retained the P-element in the genome (by eye color marker on the element, for example); set up each fly into its own vial so that its particular insertion location(s) can be studied.
- Screen the offspring (embryos) of each of these flies by using the *lacZ* gene to generate a blue product in the cells specified by the enhancer the P-element has inserted near.
- If any flies are found in which embryonic muscles turn blue, these flies are candidates for having an insertion in or near a myogenic gene.
- Isolate genomic DNA from the flies of the lines that have blue muscles.
- Use inverse PCR to amplify the sequence flanking the P-element.
- Probe genomic and cDNA libraries with that flanking sequence.
- Test any cDNA clones that hybridize with the flanking genomic DNA for hybridization to transcripts in embryonic muscle.
- Continue to probe cDNA libraries with genomic sequence until an embryonic muscle-specific (or muscle-adjacent) staining pattern is found that matches (nearly anyway) the P-element induced lacZ pattern of the line.
- Sequence that cDNA . . .

5. The answers to these questions involve finding the appropriate web site on the Internet. This can be done by using a search engine (such as Yahoo) or by entering the URL for the site required (assuming you know it) in your browser window.

(a) The multiple cloning site of pcDNA4 myc HisA includes the following restriction sites: *Hind*III, *Acc*65I, *Kpn*I, *Bam*HI, *Bst*XI, *Eco*RI, *Pst* I, *Eco*RV, *Bst*XI, *Not*I, *Xho*I, *Bst*EII, *Xba*I, *Apa*I, *Sac*II, *Bst*BI, *Age*I, and *Pme*I.

(b) The selectable markers in pcDNA4 my HisA include the ampicilliin resistance gene (for selection in ampicillin-sensitive *E. coli*) and neomycin (for selection in eukaryotic cells).

(c) This vector is about 5.1 kb. Its full sequence can be obtained at www.invitrogen.com/vectors.html

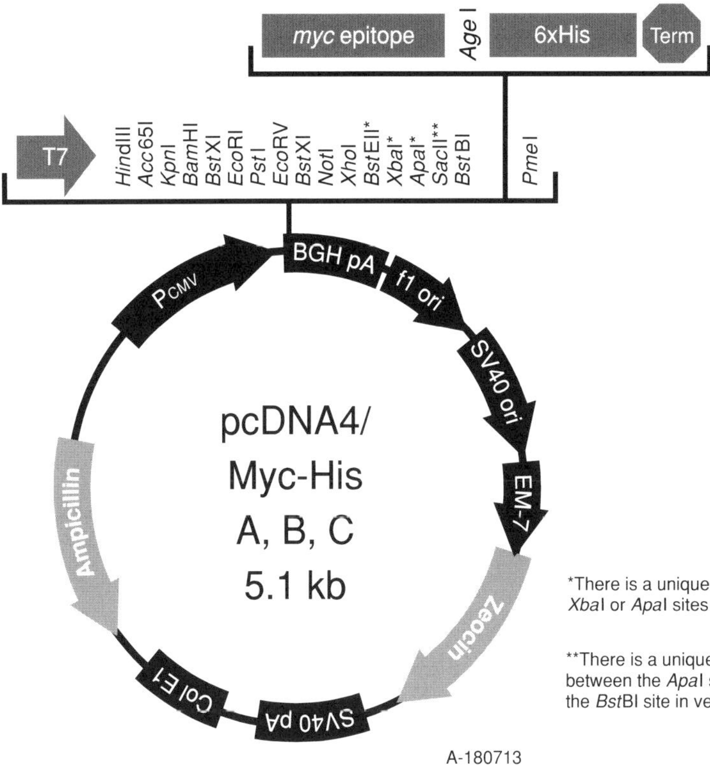

*There is a unique *Bst*EII site, but no *Xba*I or *Apa*I sites in version C.

**There is a unique *Sac*II site between the *Apa*I site and the *Bst*BI site in version B only.

5 Nucleic acid hybridization assays

Answers to MCQs

1. D 2. E 3. C 4. E 5. A 6. B 7. C 8. B 9. E 10. C

Open-ended problems

1. The degeneracy of a probe can be calculated by determining how many different codons can encode each of the amino acids required. For example, the amino acid isoleucine (I) can be encoded by three different codons, AUC, AUU and AUA. So, one would need to design three types of probe to be sure that one of the probes matched perfectly at the isoleucine codon. This one-codon probe, then, would have three-fold degeneracy. For even the shortest probes made based on peptide data, several codons must direct the probe sequence. So we determine the number of codons possible at each position and *multiply* the resulting numbers together to determine how many combinations total there will be if we want to synthesize a probe containing all possible combinations.

   For {I C G T P S Y K N L C R T} the calculation is as follows:

   $3 \times 2 \times 4 \times 4 \times 4 \times 6 \times 2 \times 2 \times 2 \times 6 \times 2 \times 6 \times 4 = 5\,308\,416$ fold degeneracy

6 PCR, DNA sequencing and *in vitro* mutagenesis

Answers to MCQs

1. E 2. A 3. B 4. E 5. E 6. A 7. B 8. E 9. C 10. C

Open-ended problems

1. One way would be to design primers as shown below, and to use them to generate a 1 × 1M product in one PCR reaction and a 2 × 2M product in a second PCR reaction. These two products could then be combined for a final reaction involving one strand from each of the two products hybridizing and being filled in by the polymerase. See *HMG2, Figure 6.20B*.

Primer 1 (forward)	5′	GTA TAC GTA CCT GGT AAG ATT	3′	Binds to antisense strand bases 1 to 21
Primer 1M (reverse)	5′	T GGT TCC GCT TAC TAA TGT A	3′	Binds to sense strand bases 553 to 534
Primer 2M (forward)	5′	T ACA TTA GTA AGC GGA ACC A	3′	Binds to antisense strand bases 534 to 553
Primer 2 (reverse)	5′	GGT ATT CAT AGA ATA GGT ACA	3′	Binds to sense strand bases 1101 to 1081.

7 Organization of the human genome

Answers to MCQs

1. D 2. B 3. A 4. E 5. E 6. E 7. D 8. E 9. E 10. E

Review

1. ***Retropseudogenes***
   cDNA-like copies of RNA that has been reverse transcribed by a reverse transcriptase (not encoded by pseudogene). Though these pseudogenes have undergone transposition before, they are not especially mobile, e.g. the human *Alu*repeat family.

   Retrotransposons
   Lack LTRs and other retrovirus-like sequences (e.g. those encoding coat proteins . . .) but encode reverse transcriptase and are thus able to catalyze their own transposition, e.g. the human and mouse LINE-1 elements.

   Retroviruses
   Viruses which integrate into the genome; possess LTRs and encode reverse transcriptase and other viral proteins. Retroviruses are able to move between cells because they can be packaged and secreted, e.g. human immunodeficiency virus.

   Retroviruses can infect host, challenge the host immune system, insertionally inactivate genes, provide sites of ectopic recombination (unequal cross over, unequal sister chromatid exchange), and promote duplication or deletion of genetic material (by facilitating intrachromosomal recombination between two retroviruses integrated near each other). The insertion of retropseudogenes near or within other genes can generate a mutation of the gene, often a complete null.

   If the element host has a restricted range (as for *Alu*), the sequence can be used as a marker for the host-species DNA in modern PCR-based analyses.

Open-ended problems

1. The DNA of each type was chosen for its overall complexity. Poly-A+Poly-U has a one base pair long repeat. Mouse satellite DNA has, on average, repeats of

several hundred base pairs. The T4 genome is approximately 2×10^5 base pairs with little repeated sequence within it. Similarly *E. coli* has little repetitive sequence in its genome of 6×10^6 base pairs. Finally, though the DNA of eukaryotes would tend to have a lot of repeat sequence, the calf DNA in this figure, the nonrepetitive fraction, has been depleted of most of its repetitive content.

In separate experiments a known initial concentration (C_o) consisting of many thousands to millions of copies of each complex double-stranded nucleic acid is sheared by sonication to fragments about 500 base pairs long. These fragments are then boiled to denature them and make them 100% dissociated or, put another way, a 0/1 fraction reassociated. Once all the DNA is single stranded, the solution in removed from the heat and monitored continuously (in the case of the poly-U+ poly-A by UV spectroscopy, or at time points (the dots defining the curves for the other DNA samples). The density of the dots reflects the number of time points at which the fraction reassociated was measured. As time passed (that is t increased), the fraction of DNA reassociated approached 1.0. Since how fast the reassociation occurred is directly related to the starting concentration of DNA fragments, the x-axis is not just time (t) but the product of initial concentration, C_o, and time (C_ot). Therefore the units of the x-axis are mole/liter · sec.

What the data tell us is that complexity of DNA mixture can be assessed by reassociation rates. Thus, presumably, one could shear genomic DNA from any species, denature it, and observe its reassociation kinetics – and from that learn about the fraction of the genome that is highly or moderately repetitive, and the fraction that is not repeated.

2. DEAD box family of proteins

   (a) to find a DEAD box-containing protein, go to GenBank (EMBL or JDB are as good) and type 'DEAD box' into the query window. This will search all records for any that contain the term 'DEAD box'. A little closer scrutiny of the records obtained will identify the bona fide DEAD box-family listings. Select the portion of the record that identifies the amino acid sequence (usually follows the identifier 'translation=' in a section called 'CDS' which identifies the coding sequence) the accession number of one such GenBank entry, whose protein sequence is listed in part below, is AJ250027.

```
MAAAAFEVPAALTTSESTMAAERAAAPVQAVEPTPASPWTQRTAHDIGGPRTRTGDVVLAEPADFESLLLS
RPVLEGLRAAGFERPSPVQLKAIPLGRCGLDLIVQAKSGTGKTCVFSTIALDSLILENYSTQILILAPTRE
IAVQIHSVITAIGIKMEGLECHVFIGGTPLSQDKTRLKKCHIAVGSPGRIKQLIELDYLNPGSIRLFILDE
ADKLLEEGSFQEHINWIYSSLPASKQMLAVSATYPEVLANALTRYMRDPTFVRLNPSDPSLIGLKQYYQVV
NSYPLAHKIFEEKTQHLQELFSKVPFNQALVFSNLHSRAQHLADILSSKGFPTECISGNMNQNQRLDAMAK
LKQFHCRVLISTDLTSRGIDAEKVNLVVNLDVPLDWETYMHRIGRAGRFGTLGLTVTYCCRGEEENMMMKI
AQKCNINLLPLPDPIPPGLMEECLNWDVEVKAAMHTYSSPTVATQSPKKQVQKLERAFQSQRTPGNQTPSP
RNTSASALSARPKHSKPKLPVKSHSECGVLEKAAPPQESGCPAQLEEQVKNSVQTSVEDSSSNSQHQAKDS
SPGSLPKIPCLSSFKVHQPSTLTFAELVDDYEHYIKEGLEKPVEIIRHYTGPEAQTGNPQNGFVRNRVSED
RAQMLVSSSQSGDSESDSDSCSSRTSSQSKGNKSYLEGSSDTQLKDTECTPVGGPLSLEQVQNGNDTPTQV
EYQEAPETQVKARHKEGANQRSKQSRRNPARRSSYRVQSEPQEESWYDCHRETTASFSDTYQDYEEYWRAY
YRAWQEYYAAASHSYYWNAQRHPSWMAAYHMNTVYLQEMMRGNQ
```

   See HMG2 for conserved features (shaded above) in the DEAD box family.

   (b) At the time of this writing, hundreds of DEAD box genes have been sequenced. The organisms found to have DEAD box genes include (but are not limited to!) the following: *Homo sapiens* (Humans); *Drosophila melanogaster* (fruit fly); *Arabidopsis thaliana* (mustard); *Saccharomyces cerevisiae* (yeast); *Mycobacterium tuberculosis* (a bacterium).

8 Human gene expression

Answers to MCQs

1. D 2. A 3. B 4. A 5. E 6. C 7. E 8. B 9. D 10. A

Open-ended problems

1. The diagram shows: (a) one promoter; donor and acceptor sites flanking all exons; (b) multiple promoters (in front of all variable exons); splice sites only after exons; (c) multiple promoters including one in front of constant domain exons; evidence of trans-splicing; (d) consensus DNA rearrangement sequences between each variable module; enhancer upstream of constant regions; recombinases in somatic cells (neurons).

(a) The first model could be investigated by searching the sequence corresponding to the shaded area for elements of a promoter. A clear promoter should be there. The regions upstream of the putative variable exons would be expected to be free of functional promoter sequences.

The model also requires that functional donor and acceptor splice sites flank all of the exons. Sequence analysis could establish this.

(b) The second model could be investigated by sequence analysis as described above. In this case, however, one would expect to find a functional promoter in front of all of the variable exons and a splice site only *after* each exon.

(c) The third model could be investigated by identifying a promoter sequence in front of the constant region exons as well as before all the variable region exons. Trans-splicing would be shown by the absense of splice intermediates containing multiple variable exons. These splice intermediates could be found, if they exist, in the heterogeneous nuclear RNA.

(d) The fourth model depends on the presence of consensus DNA rearrangement sequences between each variable module. Upstream enhancer regions would be exptected before the constant region, and recombinases would need to be found in the somatic cells in which the recombination is hypothesized to occur.

9 Instability of the human genome: mutation and DNA repair

Answers to MCQs

1. C 2. E 3. D 4. C 5. C 6. A

Open-ended problems

1. (a) Missense mutation [transversion] that abolishes translation or now encodes an *N*-terminally truncated protein.
   (b) Missense mutation [transition] that changes an internal methionine to a threonine (a non-conservative substitution).
   (c) A silent (synonymous) mutation [transition] in which a codon for leucine is changed but still encodes leucine.
   (d) A nonsense mutation [transition] that changes a codon for an internal tryptophan to a stop codon.
   (e) A triplet expansion mutation, probably by slipped strand mispairing, that increases the number of glutamines in a protein by three. This may or may not be pathogenic.
   (f) An insertion mutation that will cause a frame shift, and probably a stretch of nonsense amino acids and a C-terminal protein truncation.
   (g) A silent (synonymous) mutation [transition] in which a codon for glycine is changed but still encodes a glycine. This mutation could, however, activate a cryptic splice site, causing the mis-splicing of introns and possibly the loss of protein sequence, addition of gibberish amino acids, and/or the truncation of the protein.

2. In females, primary oocytes arrest in prophase of meiosis I until ovulation, the time between puberty and menopause when a particular primary oocyte matures to the secondary oocyte stage and is available for fertilization. Homologous chromosomes in the meiotically arrested oocytes have been paired for years before they are finally allowed to complete meiosis. This increases the likelihood of nondisjunction events significantly, and explains why there is an increase in the incidence of chromosomal problems in conceptuses with increased maternal age.

   In males, there is no arrest of primary spermatocytes so nondisjunction rates are not particularly high. But given the number of cell divisions a spermatogonium goes through to generate a constant supply of spermatozoa, the opportunity for endogenous mutations occurring during so many rounds of DNA synthesis is higher in males than females.

3. First, observe that there are two point mutations (underlined) on the paternal chromosome and that they are both now identical to the base that is present in the maternal homolog. This suggests (though, of course, doesn't prove) that gene conversion has occurred in which a maternal chromosome strand invaded the paternal chromosome at a point of breakage, and served as the template for the synthesis of DNA on a free 3′ OH and of the paternal chromosome strand. See *HMG2, Figure 9.10(C)* for one way this can occur.

4. (a) Unequal crossover.

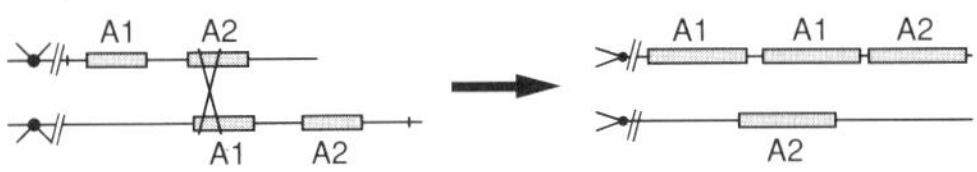

(b) Unequal sister chromatid exchange.

(c) Two-transponson-mediated intrachromatid recombination and local reintegration,

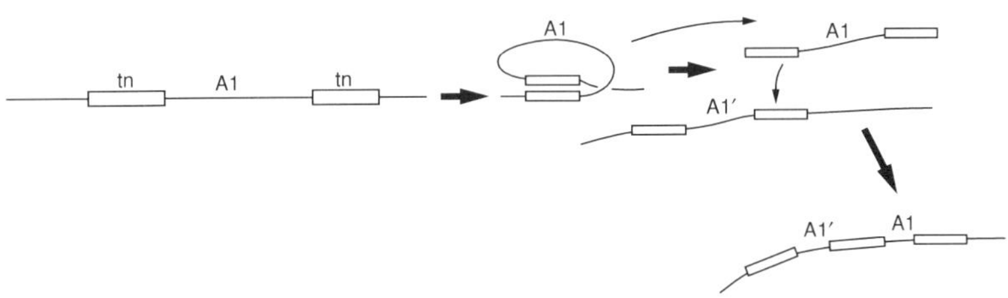

(d) X-ray-mediated double strand break followed by integration or telomeric ligation of the piece at another chromosomal location.

10 Physical and transcript mapping

Answers to MCQs

1. A 2. A 3. C 4. A 5. E 6. A 7. A 8. C

Review questions

1. EST: Expressed Sequence Tag – any DNA sequence that corresponds to an actively transcribed gene, is derived from a messenger RNA (by, for example, arbitrarily primed RT–PCR) and used to identify this specific gene among thousands of other genes. 30 000 ESTs were mapped by October 1998. The next goal is to obtain full-length cDNA sequences for all human genes.

   3p21.3 ESTs include: [see NCBI's GeneMap '99]
   A005P36 (=*ITIH4* gene)
   AI004137 (Soares_testis_NHT *Homo sapiens* cDNA = *ITIHL1* gene)
   WI-9257 (=*STK2* gene)
   C21257 (=HUMGS0002305)
   N67096 (=yz45h02.s1 *Homo sapiens* cDNA cone 286035)
   EST to *BRCA1*
   EST to *BAP1*

   STS: Sequence Tagged Site – any short track of about 200–500 bp that is unique to a given genomic DNA fragment and serves to identify that fragment among thousands of other fragments used to construct a genetic and physical map of a eukaryote genome. 30 000–50 000 STSs distributed throughout the human genome are sufficient to construct a physical STS map of the whole genome. 52 000 STSs were mapped as of October 1998.

   3p21.3 STSs include: [see NCBI's dbSTS]
   stSG48426 (RH93187)
   SHGC-53936 (RH42911) accession No. G33302
   WI-12651
   R06398
   sts-X69532
   stSG54766

   Microsatellites – a small run (usually less than 100 bp) of very short (2–4 bp) middle repetitive, tandemly arranged, highly variable DNA sequences dispersed throughout eukaryotic genomes. For example, the microsatellite sequence $(TG)_n$ is present at about 5–10 × 10^4 copies per human genome, spaced at intervals of 50–100 kb. There are estimated to be greater than 100 000 microsatellite loci in the human genome.

One microsatellite at 3p21.3 is:
D3S3726 (AFMa083ye5)

Open-ended problems

1. The gene lies somewhere within the bands 5q15 to 5q21. This is determined by recognizing that all the somatic hybrid cells that are positive for the gene (as shown by PCR) include these bands and all the somatic hybrids that are negative for the gene do *not* include these bands. The concordance provides strong evidence for this subchromosomal localization.

2. Since the DNA underlying the P1 has been sequenced already, one first needs simply to identify the specific contents of the P1 clone. Dideoxy sequencing in from the ends of the P1 vector will allow one to determine the sequence of the ends of the insert. Use that end sequence as the query sequence in a BLAST search against the *Drosophila* genome in the NCBI database. The two matches should define about 100 kb of sequence between them.

   It is likely that the open reading frames (potential genes) in this 100 kb piece will already be identified in the annotations of the sequence. These were determined using gene-finding computer programs (such as Grail) that recognize promoter sites, splice sites, and long open reading frames. Gather the names of these annotated genes to start with.

   It is also possible, however, that the set of genes identified may not be accurate or complete. To find additional genes in the insert, labeled restriction fragments of it could be used to probe developmental Northern blots. If a given fragment probe reveals one or more bands on the Northern, that suggests that the fragment corresponds to an expressed region of the DNA, and possibly is a *bona fide* gene. Any of these that are identified can be added to the list of annotated genes with the caution that these could be expressed pseudogenes, too.

   Finally, one will want to see if functional RNA or protein products are actually made to the expressed genes. This will usually be done by biochemical and genetic analysis.

11 Genetic mapping of mendelian characters

Answers to MCQs

1. D 2. C 3. B 4. B 5. E 6. C

Review questions

1. A Z value of 0 is inconclusive; there is neither support for nor exclusion of the linkage of the two alleles. More informative meioses need to be scored.

   A Z value of –3 allows one to exclude linkage at the theta values that give a Z value of –2 or less.

   Z values diverge from zero (i.e. go up to higher positive numbers or down to lower negative numbers) as the number of informative meioses increase.

   Theta is the same as recombination fraction. It can be observed (i.e. 20/337 or 20 recombinants in a total of 337 offspring of informative meioses) or assigned *a priori*.

   Map distance is actual recombination fraction × 100 (i.e. a percentage recombination between two sites).

Open-ended problems

1. $\text{PIC(A)} = 1 - [(0.1)^2 + (0.1)^2 + (0.1)^2 + (0.7)^2] - [2^*(0.1)^2(0.1)^2 + 2^*(0.1)^2(0.8)^2 + 2^*(0.1)^2(0.8)^2]$
   $= 1 - 0.660 - 0.026$
   $= 0.314$

   $\text{PIC(B)} = - [(0.3)^2 + (0.3)^2 + (0.4)^2] - [2^*(0.3)^2(0.3)^2]$
   $= 1 - 0.5 - 0.125$
   $= 0.375$

   So, though neither are very good, locus B has a higher polymorphism information content and would therefore be a better marker.

2. $Z = \log_{10} [(1 - \text{theta})^{\#NR} \times (\text{theta})^{\#R} / (0.5)^{\#T}]$ which can be rewritten as follows . . .

   $Z = NR \log_{10} (1 - \text{theta}) + R \log_{10} (\text{theta}) - T \log_{10} (0.5)$

where NR = number of nonrecombinants, R = number of recombinants and T = total number and theta = the recombination fraction (between zero and 0.5).

Here, theta will be solved for 6 values: 0, 0.1, 0.2, 0.3, 0.4, and 0.5.

NR is either 490, 458, or 350. R is either 10, 32 or 150, respectively. T = 500.

Z, then, is solved 18 times (6 theta values × 3 sets of recombination data). The results are shown in a table below and can easily be graphed with lod score on the *y* axis, theta on the *x* axis.

X1-1	NR=490, R=10
0.0	negative infinity
0.1	118.09
0.2	96.04
0.3	69.38
0.4	37.83
0.5	0.00

Z_{max} = 129.23 (at theta = 0.02)

X2-6	NR=458, R=42
0.0	negative infinity
0.1	87.56
0.2	76.77
0.3	57.61
0.4	32.19
0.5	0.00

Z_{max} = 87.88 (at theta = 0.084)

X2-6	NR=350, R=150
0.0	negative infinity
0.1	–15.50
0.2	11.75
0.3	17.87
0.4	13.18
0.5	0.00

Z_{max} = 17.87 (at theta = 0.30)

It is clear that all three markers *are* linked to the gene that causes the obese phenotype in these mice. Locus X1 is 2 map units away [between 0.94 to 3.60 map units (that is, the theta values that give a Z score of one LOD unit less than Z_{max})]. Locus X2 is 8.4 map units away (between 6.0 and 11.5 map units). And X3 is about 30 map units away (between 25.7 and 34.5 map units), though we should use a mapping function to determine the distance and range precisely.

If these had not had Z_{max} scores over 3.0, one could look for linkage at other loci, or if there seems to be a trend that suggests linkage to a locus, one could look at more informative meioses in the region. The data above *exclude* linkage within 13.5 map units of locus X2 since at theta values lower than 0.135, Z values become less than –2.

12 Genetic mapping of complex characters

Answers to MCQs

1. C (see *HMG2*, Figure 12.4). 2. D, using the Yule coefficient equation from *HMG2*, section 12.4.2. 3. E, because the correct value is 0.05/n or 0.05/100 = 0.0005 (see *HMG2*, Table 12.2). 4. A (see section 12.6, *HMG2*). 5. B (see Figure 12.1, *HMG2*)

Review questions

1. Classical disease-marker association studies involve looking for association between marker loci and disease state in large numbers of unrelated affected individuals. The transmission disequilibrium test examines parents who have one or more affected offspring and asks whether for any given marker for which the parent is heterozygous (of thousands of markers that might be tested), which allele associated with the disease state in the affected child. For unlinked genes, allelic association will be approximately 50%. For closely linked genes the association of a parents allele at the locus with the child's allele at the same locus will approach 100%.

   In classical disease-marker association studies one doesn't need to obtain multicase families or parents of affected children; in fact the studies require no family information. If an association is found, it defines a small candidate region in which to search for the candidate gene. Association studies are more powerful than linkage studies for detecting weak susceptibility alleles.

   The transmission disequilibrium test overcomes many of the problems of classical association studies, including problems of stratification. Classical association studies are also better suited to study late-onset diseases where parents may not be available for typing. Also, classical association studies are better suited for investigating environmental effects that can make a large difference in phenotype.

   The best approach to identifying susceptibility genes might begin with genome-wide screen by linkage and then, once the gene has been localized to a candidate region, to narrow in on it by linkage disequilibrium mapping.

2. There are several reasons why some of the published disease-marker associations have not been able to be reproduced.

   (a) inappropriate or insufficiently rigorous statistical test.

(b) a low relative risk for the susceptibility allele.
(c) allelic association due to reasons other than linkage disequilibrium: direct causation, natural selection, or population stratification.

Open-ended problems

1. (a) Significant linkage is established when a *p* value or lod score is obtained that would be expected to occur by chance 0.05 times in the whole genome scan (i.e. 1 000 000 tests). For this, $\alpha = 5 \times 10^{-8}$, $Z_\alpha = 5.33$, and $Z_{(1-\beta)} = -0.84$ (assuming 80% power).

(b) γ = the relative risk (RR) conferred by one copy of the susceptibility allele = ad/bc where
a = proportion of affected individuals with the allele,
b = proportion of affected individuals without the allele,
c = frequency of the allele in unaffected (control) individuals,
d = frequency of absence of allele in unaffected (control) individuals.

Here, $\gamma = 0.7 \times 0.9/0.3 \times 0.1 = 21$

(c) Frequency of susceptibility allele, $p = 0.05$. The frequency of the + allele, $q = 0.95$. So, the proportion of parent of affected children that will themselves be heterozygous for the susceptibility allele is

$h = pq(\gamma + 1)/(p\gamma + q) = 0.05 \times 0.95\ (21 + 1)/(0.05 \times 21 + 0.95) = 0.523.$

(d) $P(trX) = \gamma/(1 + \gamma) = 21/(1 + 21) = 0.95.$

(e) The mean, μ = [square root of h]$(\gamma - 1)/(\gamma + 1)$] = [s.r. (0.523)](20)/(22) = 0.657
The variance, $\sigma^2 = 1 - [h(\gamma - 1)^2/(\gamma + 1)^2] = 1 - [0.523(400)/(484)] = 0.568$
$M = (Z_a - \sigma Z_{1-b})^2/\mu^2 = (5.33 - (0.568)(-0.84))^2/(0.657)^2 = 67$
Since each parent acts independently, the required number of parent-child trios, N, is one half of M, or 34.

(f) Assumptions include the following:
(i) the relative risk of a person carrying two susceptibility alleles is γ^2
(ii) the marker used is always informative
(iii) there is no recombination between the marker and susceptibility locus.

13 Genome projects

Answers to MCQs

1. E 2. A 3. D 4. D 5. C

Review questions

1. Clone-based genomics uses cDNA and genomic libraries, and leads to the formation of a high resolution physical map. Disease-based genomics is based on the analysis of a disease or a phenotype and its inheritance pattern within a family. Disease-based genomics leads to the formation of a genetic map. Both methods use linkage analysis with the ultimate goal of eventually forming a complete map of the human genome. The advantages of clone-based genomics are that it can lead to the identification of new genes, it can be used to identify polymorphisms, and can provide a more accurate location of the genes. The advantages of disease-based genomics are that this method can lead to the identification of a gene involved in a disease, and may provide insight into the actual mutation or abnormality which is causing the disease. A genetic map may also be used as a diagnostic tool to identify an individual's predisposition to a certain disease.

2. *Haemophilus influenzae* was the first free living organism for which the entire gene sequence was determined. The gene sequence is used to understand evolutionary biology and gene functions. The association of this organism with disease also make it an important one to understand.

   Myobacterium genitalium is the smallest autonomous self-replication organism. It is used for understanding what a minimal genome should be.

   Methanococcus jannaschii has an archael genome and is used mainly for evolutionary studies.

   Escherichia coli has been studied extensively. It is used for analyses of gene function, regulation, and structure, as well as for understanding DNA replication.

   Saccharomyces cerevisiae is one of the most extensively studied eukaryotes. Many of its genes have human homologs. Many of the gene functions, such as the DNA repair mechanisms have been conserved through evolution. Its gene sequence is also used for functional and comparative analysis.

   Caenorhabditis elegans. This round worm is transparent, which makes it a good organism for conducting lineage and expression studies. The lineage of every cell in this animal is known. It is also used for studies of the nervous system.

Drosophila melanogaster is small, has a short life, and is easy to maintain in a laboratory. It has been extensively studied. Its polytene chromosomes and p-elements allow scientists to perform mutation studies. Many of the fruit-fly genes have homologous genes in humans.

Mus musculus is one of the most extensively studied mammals. Its short life and small size make it a good candidate for laboratory studies. Most mouse genes have homologous human genes.

Fugu rubripes rubripes is used for comparative studies. The puffer-fish has a compact genome with very small introns. It is used for identifying human coding sequences.

Zea mays is also used for comparative studies. This is a tetraploid plant which is used to study various aspects of Mendelian genetics.

Open-ended problems

1. Design one of the following pair of oligonucleotide primers: (–376): 5′ CCTGTCTGGAAG3′ wild-type; 5′ CCTGTCTGGAAG3′ mutant

   Affix each primer to a different location in the array of oligonucleotides (a DNA microarray). Prepare genomic DNA from the subject, cut, amplify and denature it, and add it to the array. Add DNA polymerase, buffer and fluorescently labeled ddNTPs. If the subject has the wild type allele, then hybridization will occur at the location on the chip that contains the bound wild-type primer and so that spot will be fluorescent. If the subject has the mutant allele, then the DNA in the box with the mutant primer will be fluorescent. The subject is heterozygous if the DNA in both boxes shows fluorescence.

14 Our place in the tree of life

Answers to MCQs

1. C 2. E 3. D 4. A 5. B

Review questions

1. The serial endosymbiont hypothesis suggests that aerobic eukaryotic cells evolved by a series of endocytotic events in which an anaerobic host cell took up an aerobic organism such as a eubacterium. This event is thought to have occurred with the rise of oxygen levels in the atmosphere, giving the newly formed aerobic organism a selective advantage. Evidence for this hypothesis is based on the similarities evident between the mitochondrial genome and the eubacterial genome, such as the small genome size, lack of introns, high percentage of coding DNA, and lack of repetitive DNA sequences. Polygenetic analysis of the mitochondrial genome indicates that it is specifically similar to a specific strain of bacteria. Evidence still remains to be obtained as to the type of the host organism (nucleated vs. archeon), and the timing of the endocytotic events.

   The syntrophy hypothesis suggests that there may have been a simultaneous appearance of a genome that is present in the precursors of nuclear cells and of mitochondria. It is believed that an anaerobic, hydrogen consuming, autotrophic archeon endocytosed another anaerobic, but hydrogen producing, eubacterium. This event was driven by the symbiotic metabolic association (syntrophy). The host organism lost its autotrophic mechanisms over time, and became a heterotroph, while the symbiotic organism developed the mechanism of oxidative phosphorylation. Evidence for this hypothesis includes the fact that some eukaryotes that do not have mitochondria possess metabolic enzymes similar to those of eubacteria. Other supporting evidence is the presence of hydrogen producing organelles in some anaerobic protozoa and fungi. Evidence still remains to be obtained to support the development of an oxygen consuming organelle from a hydrogen producing one.

2. It is believed that the X and Y chromosomes evolved from a pair of autosomes. The appearance of the sex determining allele on the Y chromosome led to selective pressures against recombination occurring in that region. Lack of recombination between the two chromosomes eventually led to loss of function in the Y chromosome and pressure on expression of the genes on the X chromosome. The dosage discrepancy between the two sexes selected for X-inactivation. Early in the development of the eutherians, a large translocation occurred from an autosome to the X chromosome, making those genes subject to inactivation. The X

chromosome has been highly conserved throughout mammalian evolution. The gradual loss of function in the Y chromosome, coupled with its lack of recombination and high frequency of mutations is expected to lead to the extinction of the Y chromosome. In that case, sex determination will have to be achieved through the number of X chromosomes present in an individual.

Open-ended problems

1. One mechanism for the increased number of semaphorin genes in humans may be the duplication of the entire genome. The human genome is approximately four times the size of the *C. elegans* genome. Also, single copy genes in some invertebrates have four homologous copies in humans. Such evidence points to the possibility that the invertebrate genome may have been duplicated twice during evolution. These duplications were then followed by divergence, which restored the diploid state of the organisms and diversified the duplicated genes.

   Another mechanism may have been the duplication of portions of the genome which could have led to the duplication of the semaphorin gene. The evidence for the occurrence of such events is the presence of tandemly repeated sequences.

15 Identifying human disease genes

Answers to MCQs

1. E 2. B 3. A 4. C 5. D

Review questions

1. The general approach in functional cloning is to first identify the defective gene product involved in a particular disease, finding the gene responsible for the defective product, and mapping the gene to a location on a chromosome. The hemoglobinopathies were the first human diseases to be mapped using functional cloning.

   Positional cloning does not require any knowledge of the function of the genes involved. Instead, it requires knowledge of the subchromosomal region of the gene. Given the approximate region, candidate genes in the region are identified and tested for their relation to the disease. The gene for Duchenne muscular dystrophy was mapped using positional cloning.

2. A candidate gene can be tested by the following methods

   Mutational screening. If a mutation can be found in the candidate gene in a large subset of the affected individuals, then the gene is a good candidate.

   Restoration of normal phenotype. If the reversal of a mutation in a cell can lead to the reversal of the disease phenotype, then the gene is a good candidate.

   Production of a mouse model of a disease. If a transgenic mouse with the mutation exhibits the disease, then the gene is a good candidate.

Open-ended problems

1. The scientists may first choose a model organism, such as the dog, in which some varieties exhibit an innate love for water. The water-loving trait in the one variety could be studied just as though it were a disease or other unusual trait. Linkage analysis can be performed on pedigrees of the animals which lack the water-loving trait, and may lead to chromosomal locations that contain candidate genes. Once some chromosomal locations have been identified, positional cloning, or positional candidate gene approaches can be performed to identify candidate genes. Candidate genes can be individually tested for their relation to the water-loving trait.

16 Molecular pathology

Answers to MCQs

1. C 2. D 3. A 4. D 5. D 6. D

Review questions

1. Loss of function refers to a mutation in a gene that results in a reduction or a loss of function in the gene product. Gain of function also refers to a mutation in a gene, but in this case, the gene product may function abnormally, is increased in expression or activity, or may antagonize the function of the normal product.

   Loss of function may be caused by mutations such as
   (a) deletion of the entire gene
   (b) insertion of a sequence into a gene
   (c) a mutation in the promoter region of the gene which results in reduced mRNA production
   (d) splice site mutations
   (e) frameshift mutations.

   Example: cystic fibrosis in which the CFTR protein, an ion channel, functions poorly.

   Gain of function may be caused by mutations such as
   (a) overexpression of a gene
   (b) inappropriately open ion channels
   (c) abnormal cell surface receptors resulting in overactive G proteins
   (d) mutations in parts of multimeric proteins, leading to inappropriate packaging of the protein
   (e) mutations leading to the formation of inactive dimers

   Examples: achondroplasia, Huntington's disease (due to the addition of glutamines in the middle of the huntintonin protein), and lactose tolerance (prolonged expression of lactase past infancy).

2. Epigenetic refers to the relation between two genes (different loci!) in which the expression of one gene is a prerequisite to the expression of the other. Dominant refers to the relation between two different alleles of the same gene, in which only one of the two alleles (the dominant one) is sufficient to generate the normal phenotype in a heterozygote.

3. Atavism is the inheritance of a trait seen in remote ancestors but not immediate ones. Anticipation is the phenomenon in which a specific inherited trait is

expressed with a greater severity and an earlier age of onset with each successive generation (i.e. a grandparent has a mild form of the condition, a parent has a moderate form of the condition, and a child has a severe form of the condition).

4. The classic system for describing mutations according to their phenotype is the -morph system. Amorphic mutations (also called null mutations or complete loss-of-function mutations), make no functional product. Hypomorphic (or partial loss-of-function) mutations make less functional product (lower amount and/or lower activity). Hypermorphic mutations (one kind of gain-of-function mutation) make more functional product (greater amount and/or higher activity). Antimorphic mutations (also called dominant negative mutations) interfere with the activity of the functional alleles, reducing their activity in part or in full. Finally, neomorphic mutations (another kind of gain-of-function mutation) encode products that take on a new activity not possessed by the wild-type gene product.

5. Compound heterozygotes are those individuals who have a different mutant allele at both sites for a particular gene locus (that is, the two homologous chromosomes). They might, for example, have one null allele on one homolog and one hypomorphic allele on the other. The less severe of the two alleles generally determines the level of residual function of the product, although sometimes both alleles contribute significantly to the phenotype.

Open-ended problems

1. (a) Contiguous gene syndrome is the best answer. Autosomal microdeletion is also acceptable, though less precise.
   (b) Growth retardation may be related to the absence of *GTF2I* and *RFC2*; mental retardation and inadequate spatial cognition may be related to the absence of *LIMK1* and *WSCR1*; attention deficit disorders may be due to absence of *STX1A*; cardiovascular problems and unusual facial features may be due to the absence of *ELN*. The relationship between their genotype and other phenotypes such as the highly sociable nature and the unusual musical aptitude of the children is unknown.
   (c) The presence of the repeated sequences that flank the region make the area especially susceptible to unequal cross-overs, and can lead to the deletion of the entire region in a gamete. For this reason, deletion mutations in the region would be expected to be more common.

2. (a) Allele a is a *hypomorphic loss-of-function* mutation. There is either less protein or the protein is less functional than the normal gene product.
   (b) Allele b is a *dominant negative* mutation, somehow antagonizing the wild-type gene/gene product, interfering with its normal functioning.
   (c) Mutation c is probably a deletion of many genes. The result is a *contiguous gene syndrome*.
   (d) Mutations d1 and d2 are in different gene loci so that intergenic complementation is seen, the children all being heterozygotes M/+ at both loci. Less likely, but still possible, this could be an example of intragenic complementation, where the mutations are in the same gene but the two together are capable of producing the necessary 'activity' for a normal phenotype.

(e) Mutation e is most likely an activating mutation in an *oncogene*. That is, the mutation which alters a gene product which normally regulates cell division. In this case it appears that the mutation is in a regulatory sequence that determines the quantity of oncogene product synthesized or affects the product's rate of degradation.
(f) Mutation f is in an *imprinted* region of the chromosome. In the mother's gametes it is methylated one way, in the father's another. In this case, both sexes appears to require the gene product.
(g) Allele g probably has a *nonsense mutation*. That is, though the RNA is of the expected size and abundance, the mutation has generated a new stop codon in the coding sequence that gives rise to a truncated protein. Other possibilities include a *frameshift mutation* that leads to a premature stop or the mutational *activation of a cryptic proteolytic cleavage signal*.
(h) Mutation h is probably a *gene fusion* that generates a *chimeric protein*. The protein probably has new functions that neither of its parent molecules had. Given that the heterozygous phenotype would probably not resemble the loss-of-function in either parent gene, this would be classified as a *neomorphic mutation*.
(i) Mutation i may be at an exon–intron boundary such that the transcript is not properly spliced. If an intron were retained the processed mutant transcript would be longer than the normal mature transcript. The protein would most likely be truncated due to stop codons within the kept intron.

17 Genetic testing in individuals and populations

Answers to MCQs

1. B 2. C 3. E 4. E 5. D 6. E

Review questions

1. In genetic testing, a specific gene (or its product) is tested for the presence of a mutation. Methods used include PCR (coupled with hybridization or with use of allele specific primers), or restriction fragment analysis. In gene tracking, a family history is analyzed for the inheritance pattern of specific trait. Methods used are restriction linkage analysis, and Bayesian analysis.

2. (a) Amplify the region that might contain the mutation by PCR, cut the PCR product with a restriction enzyme that will cut only the normal or mutant form, and resolve the digested PCR product on a gel.
   (b) Amplify the region that might contain the mutation by PCR, use the PCR product to probe allele specific oligonucleotides (ASOs) on a dot blot or DNA chip.
   (c) Attempt to amplify from the region that might contain the mutation using allele specific primers (the ARMS test).
   (d) Attempt to ligate two oligonucleotides together where the 3′ end of one sits on the suspected mutation site [the oligonucleotide ligation assay (OLA)].
   (e) Attempt to amplify by PCR across a translocation breakpoint. If the amplification does not succeed, it suggests the absence of the translocation.
   (f) Assess the size of an expanded repeat by amplifying across it by PCR or by checking size by Southern blotting.

3. Scanning a specific gene for mutations can be done by many methods, each with their own advantages and disadvantages. The methods include

   (a) Southern blotting using a cDNA probe.
   (b) DNA sequencing of the entire gene, or exons, splice boundaries, and promoter.
   (c) Heteroduplex gel mobility assay.
   (d) Denaturing HPLC.
   (e) Single-strand conformation polymorphism (SSCP) analysis.
   (f) Denaturing gradient gel electrophoresis (DGGE).
   (g) Dideoxy fingerprinting.

(h) Mismatch cleavage (chemical and enzymatic).
(i) Protein truncation test (PTT).
(j) Oligonucleotide arrays (hybridization and microsequencing).

4. SSCP analysis makes use of the fact that single-stranded DNA folds into a unique stable structure based on its DNA sequence, and that a mutation in the DNA sequence alters the folded structure. DNA is denatured into single strands, and the single strands are analyzed by gel electrophoresis. Different structures appear as distinct bands on the gel. SSCP is especially useful for detecting the presence of point mutations in a gene, but does not provide the position of the mutation.

5. The PTT examines the length of the protein product artificially expressed from a cDNA or large genomic exon by gel electrophoresis. Truncations in the gene product can be detected which suggest frameshift mutations, splice site mutations, and nonsense mutations. The size of truncated product can also provide a clue as to the location of the mutation.

6. ASOs are short oligonucleotides designed to match a specific allele. They may differ by a single base from the sequence that would match another allele, but that can be sufficient to render hybridization unstable if the mismatch is in the middle of the probe. Thus hybridization of PCR-amplified sequences (from individuals to be tested for a mutation) to wild-type or mutant ASOs can be used to assess genotype.

   ASOs also allow one to test for specific mutations by taking advantage of the fact that DNA polymerases, including those used in PCR, will not synthesize DNA if the 3′ hydroxyl group of the extending chain is not on the sugar of a properly base-paired base. The ARMS test uses ASOs with mismatches at their 3′ ends in this way.

7. Two PCR reactions are done, both containing one common primer, and one unique primer. The unique primer in the first reaction is specific for the wild-type allele. The unique primer in the second reaction is designed to amplify only from the mutant allele. If an individual's DNA is amplified only in the first reaction, that individual is homozygous normal. If one or other reaction show amplification, then the individual is a heterozygote. If only the second reaction shows amplification, the individual is homozygous for the mutation.

8. Restriction fragment length polymorphism (RFLP) analysis is a procedure that relies on variation in the size of digestion products of genomic DNA that has been cut with a specific restriction enzyme, resolved on a gel, blotted to a membrane, and probed with a labeled fragment that defines the region of the genome examined. If a mutation (or polymorphism) changes a restriction site of the enzyme used in the analysis, then different sized bands are seen on the blot (or film).

9. DNA microarrays or DNA chips used for genetic testing are an array of oligonucleotides (such as 20-mers) fixed to a glass surface. They can be used in reverse hybridization assays to test for mutations in the DNA (a PCR product, for example) that is used as probe.

10. DNA microarrays can be used for sequencing genes by using primers that span the whole gene or region of interest (whose sequence for the wild-type is already known) and then hybridizing PCR-amplified target DNA to the arrays. The target DNA serves as a template for the primers attached to the chip at each location. If the primer does not match the target DNA template perfectly at the 3′ end, then the DNA polymerase will not add the fluorescent dideoxynucleotide to the primer, resulting in no signal at that position in the array. In one hybridization reaction the entire gene can be sequenced.

Open-ended problems

1. Set up three PCR reactions, all containing one common primer, and each containing one unique primer. The primer in the first reaction would amplify only a wild-type allele, the primer in the second reaction would amplify only the allele with the 4-bp insertion allele, and the primer in the last reaction would amplify only the allele with the G→C mutation. If an individual's DNA is amplified only in the first reaction, that individual is homozygous normal. If one of the other reactions also shows an amplification, then the individual is a carrier.

18 Cancer genetics

Answers to MCQs

1. D 2. B 3. B 4. A 5. A

Review questions

1. Tumor suppressor genes encode products which normally inhibit cell division when conditions in the cell are not right. Oncogenes, on the other hand, encode products that normally promote cell division. Mutator gene products are responsible for correcting any errors that occur during DNA replication. Most tumor suppressor and mutator gene mutations are recessive, while most oncogenic mutations tend to be dominant.

   HRAS: an oncogene which encodes for a RAS protein. The protein product is involved in signal transduction from receptors associated with G-proteins. Mutation of this gene causes prolonged signal transduction, leading to tumor formation.

   RB1: a tumor suppressor gene which encodes the pRb protein. The protein product controls cell proliferation by binding to and inactivating specific transcription factors responsible for progressing the cell cycle.

   MSH2: a mutator gene in humans which encodes for a protein in the MutHLS error correction system. If this gene is mutated, the correction system fails, leading to random mutations accumulating in a cell's progeny. This specific mutator gene has been related to the hereditary nonpolyposis colon cancer.

2. The two hit hypothesis requires that both copies of a gene in a single cell be mutated in order for a tumor to begin developing. The mutated gene is usually a tumor suppressor (TS) gene. When both copies of the TS gene are knocked out in a cell, the cell can begin proliferating without control, leading to a tumor.

   The multistep model of carcinogenesis indicates that there are a series of successive mutations which occur in a cell, each of which provides the cell with a growth advantage, eventually leading to uncontrolled proliferation and cancer.

3. p53 is a tumor suppressor gene. Its normal function is to direct a cell towards apoptosis if oncogenic changes are detected, or to keep a cell in the G_1 phase if DNA damage is detected. Loss of p53 function will allow a cell to proceed to the S phase even if the DNA is damaged. This increases the susceptibility of the cell to becoming a cancer cell since DNA damage may have occurred in regions which may alter oncogene functions or mutator genes.

4. Normal cells replicate their DNA with high fidelity and in a highly regulated way. Normal cells also have a circumscribed set of tissues that they remain in, often only their tissue of origin.

   Tumor cells replicate their DNA with lower fidelity and have lost essential components of their replication regulating apparatus. Tumor suppressor genes have loss-of-function mutations preventing the negative regulation of improperly dividing cells. Proto-oncogenes have mutated to growth-promoting oncogenes that keep the cell constantly in the cell cycle.

   Cancer cells have the kinds of mutations tumors have in addition to those that allow them to leave their tissue of origin and invade other tissues of the body (metastasis). These genes can encode proteases that allow cancer cells to break through the basal lamina of the tissue they derive from and defend themselves from immunological attack in the tissues they arrive in.

5. Viruses can contribute to cancer in two main ways. Retroviruses can integrate into the genome near a proto-oncogene and thereby acquire exons from the proto-oncogene that will allow the virus to becoming a *transforming* retrovirus. The viral oncogene may then encode an unregulated growth promoting factor that will convert the cell possessing it into a tumor that later may become cancerous.

   Another way a retrovirus might contribute to tumors or cancer is by integrating into the genome within a tumor suppressor gene or a mutator gene, effectively inactivating it. In this way the virus is simply serving as a mutagen that damages genes normally protecting the cell from becoming tumor-forming.

6. Oncogenic mutations have been observed in many genes. These include growth factors (platelet-derived growth factor B subunit), receptors for growth factors (epidermal growth factor receptor and macrophage colony-stimulating factor receptor), signal transduction factors (*ras* and *abl*, a protein kinase), and transcription factors (*AP-1*, *myc*, and *fos*).

Open-ended problems

1. Often a pathology sample will be a mixture of normal cells, cells that have lost one copy and cells that have lost both copies of a gene. In aggregate, then, one tends to see something less than complete loss of the wild-type allele.

19 Studying human disease genes

Answers to MCQs

1. C 2. B 3. B 4. A 5. C 6. E

Review questions

1. Height is a polygenic trait. It is possible that two individuals of medium height could be carriers of different dominant or additive alleles that affect height or could both carry recessive alleles heterozygously that when homozygous increase height. This does not argue against regression to the mean for there are many people of medium height for whom the opposite will happen (i.e. a shorter child will result) and many people tall and short who will have children of average height. As a population, however, some regression to the mean will be observed.

2. Falconer's model proposes that a threshold of gene products from several loci must accumulate for a dichotomous (one of two forms) phenotype to result. If, for instance, four gene loci contribute to a discrete illness (i.e. one that you either have or don't have, not a continuum), it may be that one will be affected if one has a susceptibility allele present at only two or three of the loci.

   Mental illness may also be best explained by Falconer's model. Many polymorphic genes affect a person's (or other animal's) mental functions. It may be that one is in the 'normal' range as long as one has a few but not all the mental illness susceptibility alleles. If one has more than a critical number, however, then the threshold is reached where the individual is predisposed to be affected by a mental illness under the right circumstances (or maybe independent of circumstances). See Section 19.3 and Figure 19.3.

Open-ended problems

1. (a) This is a bias of ascertainment problem. Use the Li-Mantel correction because all affected cases in a defined population were counted, rather than just a sample.

R = the number of affected children = 28 + 2(5) = 38
S = the number of affected singletons = 28
T = the total number of children = 2(33) = 66
p = true segregation ratio = (R – S)/(T – S) = (38 – 28)/(66 – 28) = 10/38 = 0.263

(b) Determine the relative frequency of each family composition using a truncated binomial expansion of $(1/4 + 3/4)^3$ where 1/4 is the probability of an affected child when the condition is homozygous recessive *and both parents are carriers*; 3/4 is the probability of having an unaffected child. The families with only unaffected children are not ascertained by the survey, so that outcome must be removed from the binomial expansion and the expected frequencies of the other, ascertained outcomes adjusted accordingly.

p(3 affected children in family of three) = (1/4)(1/4)(1/4) = 1/64
p(2 affected, one unaffected in family of three) = 3(1/4)(1/4)(3/4) = 9/64
p(1 affected, two unaffected in family of three) = 3(1/4)(3/4)(3/4) = 27/64
p(no affected children in family of three) = (3/4)(3/4)(3/4) = 27/64 [not ascertained]

Since the last set of families will not be discovered by a comprehensive survey of individuals with the disease, the denominator is reduced by 27 for the remaining three classes. So, 1/37 of affected families of three children would be expected to have three affected children. 9/37 would be expected to have two affected children. 27/37 would be expected to have only one affected child.

20 Studying human gene structure

Answers to MCQs

1. C (see *HMG2, Figure 20.2*) 2. D (*HMG2, Table 20.1*) 3. B (see section 20.2.2 and *Figure 20.5, HMG2*) 4. E (see section 20.2.3 and *Figure 20.7, HMG2*) 5. A (see section 20.4.1 and *Figure 20.16, HMG2*)

Review questions

1. Examination of the genomic sequence produced by the Human Genome Project and the collection of expressed sequence tags (ESTs) can give one a good idea of what a given gene's cDNA will look like. The actual DNA clones of human genes can be obtained by cDNA enrichment, protein-directed cloning, location-directed cloning, homology-based cloning, and random clone characterization.

2. Obtaining full-length cDNA sequences is somewhat more tricky as the exact start of transcription, exon choice or end of transcript may not be defined by either the bioinformatic applications or by the cDNA clones one first obtains. Nuclease S1 protection and primer extension can help one identify the true start of transcription (+1).

3. Actual exon–intron boundaries can be identified by RACE PCR, nuclease S1 protection, bubble-linker PCR and inverse PCR using various cDNA clones that are obtained.

4. Computer-based applications can help analyze DNA and protein sequences in many ways. They are used extensively in sequence entry, editing, and manipulation. DNA sequences can be scanned for promoter sequences, open reading frames, splice sites, and polyadenylation sites. Predicted coding sequences can be conceptually translated. Conceptual translations (peptide sequences) can be searched for predicted sites of post-translational modification, for structural and functional motifs, and for similarity to other proteins in the protein sequence databases. The kinds of analysis that one can perform increases steadily as the number of workers in bioinformatics increases.

5. Determining when and where a gene is expressed involves either hybridization to RNA synthesized at the gene locus or detection of protein produced from the mRNA. RNA is detected usually by Northern hybridization (often a developmental Northern or a tissue-panel Northern) or by *in situ* hybridization to the

whole organism at different stages or slide-mounted specimens. The Northern blots reveal the size and relative abundance of the transcript at various stages of development and in various tissues. *In situ* hybridization reveals the precise tissue localization, the cellular localization, and the stage-specific expression of the gene. Quantitative PCR can also assess, with some precision, the amount of mRNA produced in a given set of cells.

Increasingly expression is being analyzed by microarray analysis, where RNA from a particular tissue is labeled with a fluorescent tag, RNA from a reference tissue is labeled with a fluorescent tag of another emission wavelength, and both are hybridized to a set of cDNAs that have been spotted onto a glass chip. In this way the array of genes being expressed by a tissue under two different conditions (or of two different stages) can be assessed. Serial analysis of gene expression (SAGE) can also be used to achieve a similar result.

Full expression of a protein-encoding gene does not, however, end with transcription. mRNAs must be translated and the protein delivered to the correct cellular compartment (or secreted). Protein production at different stages of development or in different tissues can be measured by Western blot, as well as by immunohistochemistry to tissue sections. Resolution can be increased by using confocal microscopy and electron microscopy. In each of these cases the gene itself can be tagged with an antibody or other binding protein, or by being labeled with an inframe fusion to a marker such as the green fluorescent protein (GFP).

6. To determine whether regulatory sequence is transcription promoting (enhancer sequence) or transcription inhibiting (inhibitory sequence), clones of the regulatory sequence are made with different sections deleted (deletion analysis). These regulatory sequences are then fused to a reporter gene (such as beta-galactosidase, luciferase, or chloramphenicol acetyl transferase). These constructs are introduced into a cell of the appropriate type (that will have the transcription factors that will bind the regulatory sequence) and the production of reporter protein is measured. Constructs containing mostly inhibitory sequence will produce less protein than those containing mostly enhancer sequence.

7. Which proteins bind to a given piece of DNA can be determined by DNaseI footprinting, gel retardation assay, or Southwestern screening. Each of these will identify sequences which bind proteins. Southwestern screening can identify the specific binding protein, as can sequence-specific affinity chromatography using the portion of the DNA known to be bound by proteins, and a nuclear extract of the appropriate cell type.

Open-ended problems

1. Obtain overlapping cDNAs.

   RACE PCR (5′ or 3′, depending on the need).

2. Nuclease S1 protection.

   Primer extension assay.

3. Search for consensus splice sites (by computer).

   Nuclease S1 protection.

4. Search for Kozak sequences (initiation of translation consensus) (by computer).

   Sequence the N terminus of the protein by Edman degradation.

5. Search for consensus glycosylation sites (NXT), phosphorylation sites (various), secretion signal sequence (about 20 hydrophobic residues at the N-terminus of the protein), and nuclear localization signals.

   Treat protein with glycosylases or phosphatases. Use antibodies to determine cellular localization pattern (nuclear, cytoplasmic, extracellular . . .).

6. Developmental Northern analysis.

   In situ hybridization to tissue sections.

   Northern analysis of transcripts from tissue before and after drug treatment.

   DNA microarray hybridization and relative transcript level quantitation.

   Northern analysis with a panel of different tissues' RNA run on the gel.

   Quantitative PCR using different tissues as source material.

   RT–PCR.

7. CAT assay using upstream sequence to drive CAT expression.

   Deletion studies using upstream sequence.

8. DNA footprinting (DNAse protection).

   Sequence affinity chromatography.

   Computer analysis of consensus sequences in the upstream region.

21 Genetic manipulation of animals

Answers to MCQs

1. C 2. E 3. E 4. E 5. D 6. A

Review questions

1. To make a transgenic mouse by oocyte injection: (a) engineer the transgene to have the necessary coding sequence, regulatory sequence, selectable markers, and regions of homology to the target sequence; (b) inject the DNA solution into the male pronucleus of a fertilized egg; and (c) transfer the injected egg into the oviducts of a pseudopregnant mouse. If integration into the male pronuclear chromosome was successful, the embryo that develops will contain the transgene in all cells.

2. To make a transgenic mouse by embryonic stem (ES) cell transplantation: (a) engineer the transgene to have the necessary coding sequence, regulatory sequence, selectable markers, and regions of homology to the target sequence, (b) introduce the DNA solution into cultured ES cells derived from the inner cell mass of an embryo obtained from a pregnant mouse; (c) inject the genetically modified ES cells into the cavity of a blastocyst from another mouse strain (with a different coat color); (d) reimplant the blastocyst into a foster mother; (e) look for chimeras (detectable by coat color variegation); and (f) mate the chimera to generate mice heterozygous for the mutation in every cell. These can be mated to generate homozygotes.

3. To make a YAC transgenic mouse: (a) isolate YACs containing gene sequences of interest (e.g. the human heavy and light chains of immunoglobulins); (b) generate YACs with large inserts; (c) fuse yeast cells spheroplasts containing the engineered YAC to mouse ES cells; and (d) make transgenics by ES cell injection into blastocysts and introduction into foster mothers.

4. The Cre-loxP system of mutagenesis allows one to inactivate a gene in a specific cell type. A mouse transgenic for the target locus, flanked by loxP sites, is mated to a mouse transgenic for the Cre gene that has been linked to a promoter with the desired cell specificity. In the resulting heterozygote for both transgenes tissue-specific expression of Cre causes tissue-specific deletion from loxP site to loxP site, resulting in the removal of the target gene that was between them.

5. Double replacement gene targeting involves a two-step mutagenesis procedure that first knocks out a specific gene using a hypoxanthine phosporibosyl transferase (HPRT)-interrupted version of the gene to knock out and a thymidine kinase (TK) gene to select against nonhomologous recombination events. After generating the knock-out strain, the HPRT-interrupted gene is replaced (again by homologous recombination) with a gene that includes a specific mutation of interest to the scientist.

6. Knock-in mutations can be made creating a replacement vector with TK (to select against nonhomologous recombination events), the targeted gene's promoter and coding sequence interrupted by the replacement gene followed by a loxP-flanked Neo gene. After homologous recombination the targeted gene is interrupted by the sequence of the replacement gene, whose expression is now driven by the targeted gene's promoter.

7. A gene knock-out is made by designing a replacement vector containing the gene interrupted by a selectable marker (often HPRT) and a terminal copy of thymidine kinase (TK, to select against nonhomologous recombinants). $HPRT^-$ ES stem cells are transformed with the construct and $HPRT^+$ ES cells are selected for. These are injected into a blastocyst which is then inserted into a pseudopregnant mouse. Progeny are then tested by Southern hybridization or PCR for the presence of a knocked out allele.

8. Gene trapping allows one to generate mutations in (unspecified) protein-encoding genes by using a reporter gene, interrupted by an intron and lacking either a functional promoter or a functional polyadenylation site. In either case the construct must insert between a promoter or polyadenylation signal for the transgene's transcript to be produced, properly spliced and polyadenylated so that the reporter protein is made and detectable. One simultaneously generates an expression pattern and a heterozygote for a mutation in the gene that normally has that expression pattern. Given the time it takes to make specific knock-out mutations in mice, gene trapping has the potential to dramatically accelerate that process for many genes that might not otherwise have been knocked out in the near future.

9. One way to clone an animal is: (a) isolate somatic cells from an accessible tissue; (b) culture the cells in serum-depleted medium; (c) some of these cells will exit the cell cycle and enter G_0; (d) inject cells into enucleated oocytes under the zona pellucida; (e) provide an electric pulse to cause fusion of the somatic cell and oocyte membrane; and (f) transfer renucleated oocyte to a foster mother.

Open-ended problems

1. To make a LIMK1 knockout in mice: (a) obtain the mouse LIMK1 coding sequence from GenBank/EMBL/DDBJ; (b) identify restriction sites that would allow the insertion of the HPRT gene behind a promoter active in ES cells; (c) insert this HPRT-interrupted LIMK1 gene into a replacement vector carrying TK; (d) insert this complete vector (cloned and purified) into $HPRT^-$ ES cells and select for $HPRT^+$ cells; (e) inject these cells into $LIMK1^+$ blastocysts; (f) insert these blasto-

cysts into pseudopregnant mice; (g) mate the progeny and test their offspring for the presence of the HPRT-interrupted LIMK1 allele by Southern hybridization or PCR; (h) mate the heterozygotes to obtain mice homozygous for the knock-out mutation. If no homozygotes are born, it may be that the mutation is homozygously lethal. Embryos can then be obtained and examined for defects that could explain the early lethality.

2. To make a mouse that has no dopamine receptor 4 (DR4) expression in the substantia nigra using the Cre-loxP system: (a) identify a regulatory sequence that directs expression in the substantia nigra specifically. This may be the most difficult part as many regulatory regions are not as specific. One would begin by searching Medline and GenBank for genes expressed in the substantia nigra and then investigating their *complete* expression patterns. If one can find no gene which is expressed only in the substantia nigra, deletion analysis of some of the best candidate regulatory regions might allow one to identify part of a regulatory region which has that specificity. That failing, one would use the regulatory sequence that is least likely to complicate the analysis of the substantia nigra DR4 loss-of-function phenotype; (b) join the substantia nigra specific promoter to the Cre gene; (c) engineer a DR4 replacement construct by inserting the DR4 gene (or an exon thereof) between the second and third loxP sites, following the beta-galactosidase marker gene that is between the first and second loxP sites; (d) introduce the DR4 replacement construct into ES cells and select for cells that produce beta-galactosidase (those that have inserted the loxP-flanked DR4 at the DR4 locus); (e) inject these successful recombinants with a transient source of Cre recombinase and select for those that lose the marker gene but retain the loxP-flanked DR4 gene; (f) inject these ES cells into blastocysts and select the progeny that transmit the loxP-flanked gene to their offspring; (g) cross the loxP-flanked gene-bearing mice with mice with the tissue-specific expression of Cre; (h) heterozygotes can be crossed to generate homozygotes. In both cases the transcription of the DR4 gene is reduced (or completely abolished) specifically in the substantia nigra; (i) study the anatomical, physiological, and behavioral differences in these mice.

22 Gene therapy and other molecular genetic-based therapies

Answers to MCQs

1. A 2. C 3. D 4. C (see section 22.5.2 and Figure 22.12, *HMG2*) 5. A (see Box 22.1 and Figure 22.1, *HMG2*)

Review

1. (a) recessive mode of inheritance (allows low dosage to have effect)
(b) loss of function mutation (can be treated simply by gene augmentation)
(c) readily accessible target tissues (easy access like blood and skin, and culturable)
(d) small coding DNA (easy to insert in delivery vector)
(e) loose control of gene expression (wide normal range)
See Table 22.4

2. (a) adenoviral
+ non-integrating, easy to scale up, large carrying capacity
– induces strong inflammatory response, acute toxicity improvements: inflammatory response can be reduced by modification or elimination of the viral genes antigenic proteins

(b) adeno-associated
+ long duration of expression in vivo, high viral titers can be produced
– small carrying capacity, difficult to purify/scale up, possible immune response improvements: inflammatory response can be reduced by modification or elimination of the viral genes antigenic proteins; more non-essential viral genes can be removed

(c) lentiviral
+ integrating for long term expression, infects non-dividing cells
– possible insertional mutagenesis issues; concern over possible 'super virus' improvements: identify ways to give insertional specificity/selectivity; eliminate all virulence genes and others not necessary for delivery of therapy.

3. A dominant negative protein inhibitor binds to the normal copy of the protein and prevents it from carrying out its normal function (or destroys it).

Antisense RNA binds to the sense mRNA, preventing it from being translated.

Catalytic RNA-based gene therapy not only binds to the RNA but directs its enzymatic cleavage (and perhaps its degradation).

4. (a) delivery of a constitutively transcribed gene therapy construct to a specifically targeted population or type of cells.
   (b) inducible using cell-type specific promoter or processing signals
   (c) inducible using externally administered drug or other treatment

Open-ended problems

- What are the specific hypotheses to be tested in the study?
- Why is gene therapy appropriate for the condition it will be applied to?
- Is there a clear need for the therapy or innovation?
- What is the relevant physiology of the condition the therapy is being directed toward?
- What are the relevant genetics of the condition?
- What organism and what genotype(s) is the therapy directed toward? Humans? An animal model? Homozygous nulls? Hypomorphs? Heterozygotes?
- What level of variation in severity of the patients' condition will be permitted (or sought out) at the beginning of the study?

- What are the characteristics of the vector to be used? Why choose a vector with these particulars? How will it be constructed?
- What regulatory sequences will be used in the expression and/or integration vector? Was a constitutive promoter used? Inducible? Tissue specific? Is it strong, medium, weak?
- What will the transformation strategy be? Why ex vivo or in vivo?
- How will the tissue specificity of the therapy be effected?

- How will the vector's ability to get into the recipient's cells be determined?
- What is the physiological halflife of the vector and product?
- What is the effective concentration of the gene/gene product? How will it be obtained in the patient?
- What is the product's bioavailability and clearance rate?
- What is the longevity of the gene-expressing cells?

- What safety measures will be taken to protect the patients/subjects?
- What immunological responses can be exprected? What if any measures were taken to address these?
- How will the patients/subjects be identified? Will the study be double blind?
- What other aspects of the patients' health will assessed?

- How many individuals will be studied?
- What control groups will be used?
- Will any other medications that might influence the recovery of the individuals be taken concurrently? How will that affect the interpretation of the results?
- What will the measured variables be?

- How will the efficacy of the therapy be defined, measured and represented?
- When and how often will the efficacy of the treatment be tested?

- How many measurements will be used to establish 'baseline'? What other physiological parameters will be measured throughout the study?
- How long are the gene therapy results expected to last?
- Can the gene therapy be given repeatedly over the years? Will it need to be?